Can't physics be simple?

Maciej B. Szymanski

Truth is ever to be found in simplicity, and not in the multiplicity and confusion of things.
– Isaac Newton

Canadian Cataloguing in Publication Data
Library and Archives Canada Cataloguing in Publication

Can't physics be simple?
Szymanski, Maciej B., author

First Edition published August 2023

Includes bibliographical references.
ISBN 978-0-9951680-2-2 (Canada)

1. Physics. 2. Laws of nature. 3. Gravitation 4. Material energy. 5. Matter 6. Force of inertia. 7. Heat. 8. Speed of light. 9. Infinite universe. I. Title.

Library of Congress Control Number: 2023908662
ISBN Paperback 978-3-2929-4578-5 (USA)
ISBN Hardback 978-3-7274-7483-5 (USA)

For permission and other information email
maciej.b.szymanski@gmail.com.

Published by Maciej B. Szymanski

To my grandsons
Aiden and Tristen

CONTENTS

ABBREVIATIONS

CMB	cosmic microwave background
CPP	current paradigm of physics
FLH	fundamental law hypothesis
FLP	fundamental law of physics (law of nature)
g-object	gravitational object
G-P field	gravitational-potential field
G-P-FH	gravitational-potential-free horizon
G-P-FR	gravitational-potential-free region of the universe
G-P-FS	gravitational-potential-free space
GR	general theory of relativity
KE	kinetic energy
QT	quantum theory
PE	gravitational-potential energy
SLT	second law of thermodynamics
SR	special theory of relativity
VO&VC	very old and very cold stars/galaxies

MOST USED SYMBOLS

t	[T]	time
d	[L]	distance (space)
a	$[LT^{-2}]$	acceleration
v	$[L^3]$	volume
c	$[LT^{-1}]$	speed of the photon
c^U	$[LT^{-1}]$	the speed of the motion of energy U_{fluid} (the speed of gravitational wave) = the speed of the photon c
V	$[L^3]$	the range of gravitational interaction = the volume of gravitational-potential (G-P) field = the amount of gravitational potential
m	[M]	mass (matter)
F_g^δ	$[MLT^{-2}]$	direct force of gravity
F_g	$[MLT^{-2}]$	true force of gravity
F_i	$[MLT^{-2}]$	real force of inertia
F_n	$[MLT^{-2}]$	antigravity force
F_D	$[MLT^{-2}]$	force on photon due to gradient D
F_e	$[MLT^{-2}]$	electric force
F_l	$[MLT^{-2}]$	spin force
υ	$[LT^{-1}]$	speed
G	$[L^3M^{-1}T^{-2}]$	Newton's gravitational constant

K	[1]	function introduced to account for a finite range of gravity in direct gravitational interactions
E	$[ML^2T^{-2}]$	ability to do an amount of work; applies to KE and PE
U	[M]	material energy U that fills the entire universe, including the atomic space
D^U		density of material energy U
D	$[ML^{-3}]$	average energy density of mass (energy U) in G-P field
U^{eq}	[M]	equilibrium energy U; its amount reflects a degree of equilibrium of a system of interacting objects
U_{solid}	[M]	energy U the cores of fundamental particles (electron, proton) comprise
U_{photon}	[M]	a quantum of energy U_{fluid} the photon comprises
U_{fluid}	[M]	energy U that fills the entire universe between the cores of U_{solid} particles; it includes energy U_{photon}

INTENDED PURPOSE AND THE AUDIENCE

The purpose of this book is to suggest and discuss the possibility that nature, which means physics at the most fundamental level, is extraordinarily simple. The suggestion is presented in the form of a conceptual theory of the fundamental physics. The discussion covers specific topics, primarily gravitation-related, which have been selected to support the notion of the simplicity of nature. The proposed theory is called the Energy Equilibrium theory (the EE theory). Its central philosophical feature is the dehumanization of nature. Its central physics features are a finite range of gravity and material energy (matter) that fills the entire universe.

The ensuing discussion concerns purely theoretical physics that is thought to be different from the applied (mathematical) physics. The latter seeks to construct *descriptive-predictive mathematical models* of physical effects. The former , on the other hand, seeks to provide *the explanations of physical effects at the fundamental level.*

The discussions presented in this book are nonmathematical, but might not be simple for a layperson to follow. While the ideas put forward here can easily be understood and appreciated by a high-school graduate, it would require a significant effort unless s/he is provided with a guidance by a physics teacher. Without that guidance, the graduate would have to put a considerable effort into researching textbooks, scientific papers; university, Wikipedia, NASA and similar web-sites; to learn about the various aspects of physics discussed in this book. On the other hand, the following discussions should be easy to grasp for a senior university/college student in physics, philosophy of science, or engineering.

The required study effort aside, the intended audience includes those who are anxious to acquire a plain, observation-justified no-nonsense knowledge of how nature could be running the universe, which means that it includes almost everybody.

PREFACE

Similar to the ancient Greeks who humanized their Gods, modern physicists humanized nature. Most physicists believe or implicitly assume that nature knows and applies mathematics to solve the equations that are formulated by physicists to state physical laws. Physical laws are mathematical models of physical effects such as general relativity, standard model, the equations of Maxwell's electromagnetic theory, etc. Physicists also believe that nature is capable of measuring the amounts of physical quantities (e.g., the amounts of mass, charge, space, and time). Moreover, they believe that nature makes decisions concerning the strength of a force generated in a physical interaction. We know, on the other hand, that doing mathematics, measuring quantities, making decisions, and similar abilities are the attributes of the human mind. What we cannot possibly know is if nature really knows mathematics, measures quantities, makes decisions, etc., just like the ancient Greeks couldn't know if Aphrodite was really beautiful or if Zeus really cheated on Hera. By supposing that nature possesses those abilities, physicists humanize her. In this book, I will dehumanize nature by denying her the abilities to do math, measure quantities, remember, make decisions, etc.

Dehumanized nature is simple in extreme. That simplicity leads to rational explanations of some of the most outstanding mysteries in physics. Those include the role of human space and human time in physics, the weakness of the force of gravity, the direction of the force of gravity, the nonfictitiousness of the force of inertia, the absolute frame of reference, the action-at-a-distance mode of interaction, the equivalence principle, the speed of interaction, the potential energy ↔ kinetic energy conversion, the origin and the substance of gravitational waves, the absolute zero temperature, the substance of heat, the nucleus-electron interaction, the sun-planet interaction, the gravitational-time dilation, the constancy of the speed of light, the aberration of light, the stellar coronae, the quantum entanglement, and the double-slit experiment. The long-time well-established experience strongly suggests that explaining all those mysteries is impossible if relying on the current paradigm of physics.

To escape that impossibility, I will define and follow a parallel paradigm, which is put forward *to explain* rather than *describe-predict* physical phenomena. For the purpose of this discussion, the current paradigm of physics was suitably stated by Richard Feynman in his physics lectures: "The law is called the conservation of energy. It states that there is a certain quantity, which we call energy, that does not change in the manifold changes which nature undergoes. That is a most abstract idea, because it is a mathematical principle..."[1] Feynman's statement actually integrates two ideas. First, energy is an immaterial (mathematical, abstract) physical entity. Second, the laws of nature are executed according to some mathematical principles. In the following, those two ideas will be disallowed and replaced with opposite ideas: The executions of the laws of nature are nonmathematical, and energy as a material substance does exist. The two opposite ideas constitute the basis of the dehumanized-nature paradigm of physics. They arise from the consideration of a hypothesis concerning the existence of laws of nature discussed in sections 1 and 2, and an interpretation of the results of the LIGO experiment discussed in section 22.

It appears fair to suggest that the current paradigm of physics was inspired by the Pythagorean school of thought and the very title of Newton's *Philosophiae Naturalis Principia Mathematica*. However, there seems to be no indication that Newton believed in mathematics existing outside the human mind. He apparently had no interest in a paradigm that would allow for explaining physical interactions beyond their mathematical descriptions, as he clearly stated, "I frame no hypotheses."[2]

The theory put forward in this book is based, in essence, on the consideration of the fundamental laws of physics (laws of nature). Fundamentals of the theory are discussed in sections 1 through 5 and 22 through 26. Gravitational interactions are examined in sections 8 through 20 after three basic hypotheses—concerning the existence of the laws of nature (fundamental laws of physics), the range of gravitational interaction, and the dependence of that range on mass-energy content—are introduced in section 1. In section 9, a modified Newton's law of gravitation is derived. That derivation is not intended to reinvent Newton's law. Its purpose is to demonstrate that the structure of Newton's law—a law that has been proven highly successful—appears to be consistent with the hypotheses underlying the proposed theory. Most outstanding implications of the three basic hypotheses are discussed in chapter III. The fourth hypothesis, introduced in section 22, concerns the existence of physically-real (nonabstract) material energy. New interpretations of the concepts of heat energy and the electric force are suggested in chapters V and VI, respectively. The suggested interpretation of electric force is presented on a tentative basis. That presentation is intended to describe a possible approach to the explanation of the electric force effect, and it doesn't affect the remainder of the discussions presented in this book. Some widely-known physics mysteries that involve the existence of material energy are explained in chapter VII. The speed of the photon and the speed of the motion of gravitational wave are examined in chapter VIII. Evidence in support of the proposed theory is briefly outlined in chapter IX. In chapter X, the principal features of a universe that is infinite in both space and time, which emerge from the consideration of the proposed EE theory, are discussed.

Parts of the theory put forward in this book were published between 2010 and 2018.[3,4,5,6] In those publications, however, I didn't realize the issue of material energy, which is a key notion discussed in this book. Until about 2018, I hadn't believed in the existence of gravitational waves as a physical reality which, as the LIGO experiment proved in 2016, were physically real. Now, I believe that the fundamental physics and the universe would be unexplainable without accounting for the LIGO results.

CHAPTER I – BACKGROUND

1 Basic hypotheses

The EE theory discussed throughout this book is founded on the three basic hypotheses put forward in this section, and a fourth hypothesis to be introduced in section 22.

The first of those hypotheses states that *laws of nature do exist independent from the human mind*. That hypothesis is referred to as the fundamental law hypothesis (FLH).

The second hypothesis is the only possible resolution of the following gravity paradox: The range of gravitational interaction normally is assumed to be infinite even though the amount of mass which, in some way, relates to the quantity of gravitational effect, is finite. As the mass of an object cannot be infinite, the only way to resolve that paradox is to suggest that *the range of gravity (the range of gravitational interaction) is finite*. The importance of the premise of a finite range of gravity will be discussed in section 12.

The third hypothesis, which augments the second hypothesis is: *The range of gravity of an object is proportional to its intrinsic mass-energy (mass) content.*

2 Fundamental law hypothesis

In the context of FLH, the word "fundamental" means that a law of nature—which is a fundamental law of physics (FLP)—cannot have underlying principles. That is, each FLP in itself constitutes a fundamental, irreducible principle. Therefore, an FLP must have no underlying machineries as any machinery operates according to some principles. Furthermore, an FLP cannot be executed based on the results of measurements for nature doesn't own measuring devices (machineries) such as clocks, rulers, computers, weight scales, thermometers, etc. That also means that nature possesses no concepts of space and time for she is unable to measure and thus detect space and time. That is a direct consequence of FLH.

Thus, the execution of an FLP by nature cannot account for space and time. Furthermore, the execution of an FLP (a law of nature) has to be nonquantitative for a realization of a quantity by nature would contravene FLH. That is for nature is unable to measure quantities. Also, an FLP cannot be executed based on a constant, understood to be a quantity of a constant value built into the fabric of nature.

The execution of an FLP cannot be based on nature's memory for she cannot have a memory as any memory must involve time. Having no memory means that nature executes her laws in instants of time for she cannot know the present from the past and from the future. Having no memory also means that nature is unable to make decisions. That is for making a decision involves selecting this or that, while she is unable to remember and thus realize this and that. To put that another way, no decision-making can be a part of the execution of an FLP as any decision making would require the FLP to have some built-in principles. In particular, nature making a decision on the strength (the quantity) of an interaction to be generated in the execution of an FLP would contravene FLH. FLH further requires that FLPs are executed always in the same way, which means that FLPs (laws of nature) are absolute and exact regardless of any circumstance such as time, location, or a specific physical setting. Otherwise, an FLP would have to have built-in principles and machineries necessary to measure circumstances, based on which nature would be able to make a decision to execute approximately different actions under the same circumstance (say, the same initial and boundary conditions). Furthermore, FLPs must be nonmathematical for the laws of nature cannot incorporate any, including mathematical principles. Those constraints on FLPs present consequences of the statement of FLH. As there are no principles underlying FLPs, *chances are good that we will never know how FLPs are executed.*

FLH dehumanizes nature since the above-stated constraints on the executions of FLPs present the attributes of the human mind: The physicist observes, remembers, measures quantities, makes decisions, does mathematics, isn't absolute and exact, etc., while FLH disallows nature to possess those abilities. As discussed, a salient implication of FLH is that the execution of an FLP has to be nonquantitative, which means that the execution of an FLP in itself cannot result in a quantitative (observable) effect. That implication will be discussed in more detail in the next section.

It appears rational to suggest that in the derivation of a *physical law* (a descriptive-predictive mathematical model developed by a physicist), the following corollary to FLH, which will be called "the FLH corollary," should be used as a touchstone: *A successful physical law; which means that the law is logical, complete as intended, and highly accurate; should not only be consistent with laws of nature but, also, should exclude the possibility that nature possesses any attributes of the human mind.* That is, nature should not be allowed, within the realm of a physical law, to do things she cannot possibly do. Said another way, the structure of physical law should be such that it makes an explicit allowance for nature's inabilities. Concerning this issue, there is an often-quoted maxim: "Whatever can happen will happen." In view of the FLH corollary, the following maxim appears to be at least as pertinent: "Whatever cannot possibly happen mustn't be allowed to happen." The latter maxim is of greatest importance. For instance, it translates into: "one mustn't postulate laws of nature at the classical and quantum (subatomic) levels to be different for nature cannot own different sets of laws dependent on space and/or time."

The leading theories of the current physics, general relativity (GR) and, particularly, quantum theory (QT), include ideas that are often called weird. That appears to be misleading, which will be argued for in section 59. For the time being, note that there seems to be nothing wrong with a physical law looking "weird." The objective of a physical law isn't to look ordinary or normal. The objective is that a physical law is logical, comprehensive as intended, highly accurate, and in agreement with the experiments and well-established/accepted laws of nature. It matters not if it looks weird or not. Calling a physical law weird is a humanization of nature for *the adjective "weird" means strange, startling and uncommon from the perspective of the human mind.* Nature has no concept of weirdness. In short, it seems that a physical law can be as strange, startling and uncommon (weird) as it comes provided that it accomplishes its objective.

The normal view is that physics at the classical (large scale) level is different from physics at the quantum (infinitesimal) level. That view contravenes FLH as pointed out when discussing the FLH corollary. First, FLH in no way implies that an FLP applicable at the classical level should or could be different from that applicable at the quantum level. Second, according to FLH, nature is unable to own different laws depending on space (distance) and time for she is neither able to realize space and time nor make decisions. To say that another way, as nature doesn't realize space and time, she cannot possibly know if it is hours and meters or nanoseconds and angstroms out there associated with a specific physical effect she is attending. Nature is dehumanized. As a result, no different *laws of nature* (FLPs) at the quantum and classical levels are expected to exist. Granted, *different* physical laws (mathematical models) typically are required to describe physical effects at the quantum and the classical levels. However, those are intended *to describe* rather than *to explain* physical effects.

At first glance, it may seem that FLH, being a philosophical notion, is highly unlikely to yield pragmatic results. The ensuing discussion is intended to demonstrate otherwise. In particular, the explanations of a number of well-known mysteries in physics—those are listed in Preface—are suggested from the consideration of FLH. It needs to be acknowledged that researching physics based on philosophical notions has been strongly discredited by some distinguished physicists (e.g., S. Weinberg[7], S. Hawking[8]). Yet, other leading physicists believe philosophy to be an essential tool in researching physics (e.g., L. Smolin[9], C. Rovelli[10]).

3 Laws of nature (fundamental laws of physics)

FLPs can be identified based on the assertion that they have no underlying principles, and do exist independent from the human mind. Concerning this discussion, the following laws of nature (FLPs) discovered by physicists, and physical laws formulated by physicists, need to be considered.

I Law of equilibrium (second law of thermodynamics [SLT]): *Nature spontaneously eradicates nonequilibrium between physical entities, if the entities are connected by a mediating agent.* Physical entities are material, therefore, physically real objects. This law states that the execution of the SLT brings matter toward a state of equilibrium. It follows that equilibrium is nature's preferred state of matter.

Ia Law of energy-equilibrium: *Bringing a system of interacting objects toward a greater or lesser degree of equilibrium requires a release or absorption of energy, respectively. That energy will be called "equilibrium energy" and labelled* U^{eq}.

The amount of equilibrium energy U^{eq} contained in a system of objects reflects the degree of equilibrium in the system. The concept of material energy will be discussed in detail in section 22. As will be explained in section 27, law Ia also states that the potential energy (PE) contained in a system of interacting objects decreases or increases as the degree of the equilibrium of a system increases or decreases, respectively.

The well-known manifestations of the execution of law Ia are chemical synthesis and disassociation reactions, as well as phase transitions, in which heat ("energy") is released or absorbed. The relation between the common concept of heat and material energy will be discussed in section 28.

II Law of gravitational interactions: *Mass-energy objects have the potential to interact in an observable way if connected by a mediating agent.* For the potential to be activated, inflicting a conflict with the nonquantitative execution of this law is necessary. The concept of a conflict will be explained later in this section. Law II presents a specific aspect of the SLT.

III Law of electric interactions: *Charged particles (charges) can interact in an observable way if connected by a mediating agent.* Law II presents a specific aspect of the SLT.

IV Law of inertia (Newton's first law of motion): *An object is at rest or remains in uniform motion along a straight line unless acted upon by a net (unbalanced) external force.* Herein, an "unbalanced force" means a total force acting on a single-mass object that appears to be unbalanced as the influence of the force of inertia is neglected.

V Law of force-acceleration (Newton's second law of motion): *The acceleration of an object equals the net force acting on the object divided by its mass-energy.*

VI Law of action-reaction (Newton's third law of motion): *For every action, there is an equal and opposite reaction.*

VII Energy-conservation law: *Material energy cannot be created or destroyed.* The use of the word "material" will be clarified in section 22.

VIII Momentum(motion)-conservation law: *Momentum(motion) contained in a closed system is conserved.* That law will be elaborated on in section 26. Note that law VIII differs from the material-energy conservation law. The former applies to a closed system, while the latter applies to any point in space and time, and to the universe as a whole.

As pointed out in section 2, the execution of an FLP has to be nonquantitative, consistent with FLH. On the other hand, if two mass objects gravitationally interact, for instance, if an apple is falling toward the earth, the motions of the apple and the earth—which are quantitative effects—can be observed and measured. Thus, the effect of the execution of law II is clearly quantitative. It follows that beside nonquantitative execution of law II, there must be "something else" involved, which results in a quantitative effect. That supposition is consistent with the action-reaction law: The motion of an object, which represents a quantitative effect, is an action or reaction. According to the action-reaction law, the "something else" is expected to be a matching while opposite reaction or action. The existence of "something else" means that quantitative physical effects result from nonquantitative execution of an FLP and other physical influences that are the "something else." *Those influences must be interferences (conflicts) between the execution of an FLP and the executions of other, conflicting FLPs, which can also mean conflicts between the execution of an FLP and interfering (conflicting) matter*. The strong word "must" used above reflects the fact that, as far as we know, there is nothing else out there except for nature executing her laws and matter.[a] The idea of a conflict—which, in essence, is a boundary condition—suggested above is a key to appreciating the execution of FLPs, which also means a key to theoretical physics.[b]

Law I refers to equilibrium. It will be convenient to think of *progressing toward an equilibrium state of a system of objects forced by the execution of a law of nature as progressing toward a no-observable-interaction (no-motion) state*. No-motion state

---

[a] It will be suggested in section 7 that there actually is something else out there, named gravitational-potential field. That field comprises immaterial information that doesn't conflict or interact with anything while facilitates interactions.

[b] The concept of a conflict with the execution of an FLP can be difficult to appreciate. As an example of such a conflict, let a rope on which Newton's bucket is hanging go unwinding. That inflicts a conflict and the bucket quantitatively interacts with distant masses, or perhaps with something else. The interaction can be observed. Before the rope went unwinding—that is, before a conflict was inflicted—the bucket and distant masses (or perhaps the bucket and something else) were there and interacted nonquantitatively, without an observable effect.

means that no matter in a system is in net motion. Quantities in the system don't change. The system is in equilibrium for the observable execution of a law of nature ceases, that is, nature stops doing her work. (Work by nature will be discussed in section 25.) For instance, an apple hanging on a tree is in a no-motion state with respect to the earth for it is a part of the earth. They form a single-mass object. Law II isn't executed in an observable way. As an apple is falling toward the earth, it progresses toward a no-motion (equilibrium) state. After the apple collides with the earth, it is in equilibrium with the earth for they form a single-mass object again. Law II has ceased to be executed in an observable way for there is no motion. In another scenario, if two objects get separated by such a large distance that they no longer interact, the execution of law II ceases as well. It will be important to keep in mind that *nature's preferred state of matter, referred to in the statement of law I, is a no-motion state.*

Laws I, Ia, II, III and VI (the "action laws") address the motion of matter caused by the executions of FLPs.

Laws IV and V refer to a force. A force emerges as a result of an interaction. The amount of interaction is the strength of a force. The strength of a force depends on the amount of conflict only rather than on a decision made by nature. Note that laws IV and V are not considered to be true laws of nature (FLPs). They have been listed here to clarify later the status of Newton's laws of motion and gravitation.

The preceding discussion implies that physical constants used in the formulations of physical laws, which represent quantitative properties of interactions, emerge as a result of conflicts between the executions of FLPs and matter. Therefore, physical constants cannot be calculated from theory for FLPs are nonquantitative. Their values have to be determined from experiments. As such, those values should be considered approximate. (That will be discussed further in section 54.)

The statements of FLPs suggest that there are two fundamental *action laws*, law I (the SLT) and law VI (the action-reaction law). Both those laws are about equilibrium. Laws II and III present specific aspects of the SLT concerning nonequilibrium in the distribution of mass and charge. The action-reaction law states that all matter must be in equilibrium. In simple terms, an object subject to a force must also be subject to an equal while acting-in-the-opposite-direction force. There can be no "unbalanced" forces from the perspective of nature.

Concerning *conservation laws*, there is only one fundamental conservation law, the general statement of which is, "nature has no hidden sources of, or storage bins for physical entities," which is consistent with FLH's implication that nature is unable to make a decision to create or not to create (destroy) a physical entity.

In summary, FLH implies nature to be as simple as binary "0 or 1," while the human mind appears to be the most complex physical system known. It will be important to keep that in mind.

4 Space and time

It follows from FLH that the laws of nature cannot incorporate space and time for nature is unable to realize those concepts. She owns no rulers and clocks, has no memory, and is unable to make measurements (section 2). In other words, laws of nature cannot incorporate space and time as constituents. The question therefore arises: How the concepts of space and time perceived by the human mind—including the space and time that are incorporated in physical laws—can be allied with nature executing her laws?

According to Aristotle, time is change.[11] It appears appropriate to suggest that space is change as well. There are infinite numbers of independent *physical times* and independent *physical spaces*. That can be appreciated by considering an apple falling from a tree and a person driving to work. The distance (space) and time that it takes an apple to fall to the ground and a person to arrive at work are entirely independent. As an apple starts falling from a tree, physical time and physical space are born. When the apple hits the ground, physical time and physical space cease to exist for the change ceases to exist. Physical space and physical time can be observed and measured. Consequently, while immaterial and not realized by nature, they have to be considered physically real. To summarize, from the perspective of FLH, *physical space and physical time are emergent, while physically real properties of interactions between matter that nature doesn't realize.*

Concerning the concepts of *human time* and *human space*, two most remarkable traits of the human mind need to be appreciated. First, the ability to identify many unrelated physical times and correlate those thus constructing a single, continuous human time. Second, the ability to identify many unrelated physical spaces and correlate those thus arriving at a single, continuous human space. As human time and human space exist in the human mind only, they aren't physically real. Yet, to explain the world, human space and human time can be considered physically real for explaining the world is, and can be done in the human mind only. In support of that notion, it needs to be remembered that human space and human time are direct derivatives of physical spaces and physical times which, while emergent, are physically real. Nature realizes neither physical space nor physical time, which means that space and time cannot be incorporated in the laws of nature. That time should be disregarded in physics at the fundamental level has been suggested by many physicists (see, e.g., Barbour[12]). Rovelli, for instance, suggests, "...in a fundamental description of nature we must 'forget time' ...".[13]

As human space and human time are constructed in the human mind from the observations of physical spaces and physical times, it is impossible to conclude that either human space or human time is finite. That is for neither a finite space nor a finite time has ever been observed. Furthermore, as nature is unable to realize space and time, she is unable to incorporate any spatial or temporal restrictions on herself and her laws. Note that there is no hint of any restrictions, spatial or temporal, on the execution of nature's laws in the statement of FLH. It is therefore suggested that *nature (the universe) should be considered infinite in human space and in human time*. Otherwise, nature would be humanized.

Nature cannot perceive the universe the way the physicist does for she is nothing more but a set of nature laws and matter. She has been dehumanized. Nature doesn't know that space and time, and the universe exist. As she is unable to realize space and time, she is expected to execute her laws in *instants of time* and in *instants of space*. Instants of space are directions in space. There are an infinite number of time instants in human time and infinite number of space instants in human space.

Nature operating in the directions in space clarifies the meaning of the phrase "if connected" used in the statements of fundamental laws I, II and III (section 3). To be able to interact, objects have to be connected by a mediating agent that continuously extends along the straight line (*along a direction*) between the objects for nature operates in the directions of space.

It follows that for the universe to exist as we perceive it, it has to comprise two physical realities: 1- matter and the laws of nature—owned by nature, and 2- space and time—owned by the human mind. Those two realities appear to be inherently related in our perception of the universe. They suggest the existence of three fundamental physical dimensions, [M] - mass/matter, [L] - space, [T] - time.

5 Human senses

The common view is that the human mind owns five basic senses: sight, hearing, touch, smell, and taste. That view, it seems, needs an enhanced consideration such that the role of physicist's senses in comprehending nature can be better appreciated.

The sense of time, which is the sense of change according to Aristotle, isn't included in the list of basic senses. Yet, something we observe may be changing, and we can sense that change (time). The tone of voice could be changing, and we can sense that change via the sense of hearing. Thus, it isn't only that we can hear, we also can discern tones of voice varying with time (changes). The smoothness of a surface may be changing and we can sense that change via the sense of touch. The smell or taste of a substance may be changing, and we can sense those changes. Suppose a physicist is looking at her/his backyard. The day is windless. At first glance, s/he cannot sense time for s/he senses no change. Yet, owing to physicist's inherent sense of time, s/he does sense time. First, the physicist thought of the next dancing lesson, then s/he thought of yesterday's supper. Time had passed between the two thoughts. The sense of time and the underlying memory of events are inherent attributes of the physicist mind. To the contrary, nature has neither memory nor a sense of time. As a result, she operates (runs the universe) in instants of time.

The physicist is looking at her/his backyard again. Via the sense of sight, the physicist experiences *the sense of space* because s/he senses distances between trees. It seems likely that both the sense of time and the sense of space have developed in response to the instinct of survival. Say, a physicist is driving a car at a steady speed. Another car, ahead of the physicist, moves in the same direction at a lower speed. The physicist slows down in response to the instinct of survival (in response to danger). The physicist slows down for s/he sees the car ahead increasing in size. The memory makes the physicist aware that a growing object presents a danger.[c] The car ahead is growing in two directions of space. The physicist also senses a third direction; along the line between the two cars. Thus, s/he perceives a 3-directional space. *That, however, is an illusion only* because space comprises an infinite number of directions (see section 4). Assuming space to be a 3-directional, or n-directional physically-real medium means simply a humanization of nature, which is indispensable to doing applied (mathematical) physics. A physicist needs three directions to describe the location of an object in mathematics language. "Three directions" isn't a byword here. The byword is the word "needs." Certainly, *nature doesn't have any need to locate objects*. The 3-directional/n-directional space exists in the mind of the applied physicist only who cannot function without it.

Important to the appreciation of the sense of 3-directional space is to realize how the instinct of survival has affected the human eye. Suppose the human eye lens is flat. A physicist having flat-eye lenses wouldn't duck as a stone is thrown at her/him. That is for s/he would see just a speck of matter that isn't moving toward her/him since the speck isn't growing in size. S/he would see a direction (i.e., an instant) of space, just like nature does. S/he would sense no danger. Sensing 3-directional space is an attribute of the human mind. The actual, lens-like shape of the human eye leads to the illusion of continuous space. Nature with her flat-lens eyes is unable to execute her laws in relation to continuous space. *She executes her laws in separate instants (directions) of space*. That signifies the enormous power of nature. She simultaneously executes her laws in trillions of directions, in each instant of time. Suppose two copper cubes of different temperatures are brought to touch. Nature will be eradicating the nonequilibrium in the distribution of temperature separately in respect to each set of two molecules that form the two cubes. That means that she will simultaneously and independently execute the SLT in trillions of directions. The clever physicist, on the other hand, will invent

[c] The memory of a danger can be acquired, or coded in genes.

Fourier's law for heat conduction to describe the eradication of temperature nonequilibrium in three directions of space.

The human mind memory may relate the illusion of continuous time to the instinct of survival. Memory is necessary to survive. It couldn't exist without the sense of continuous time, which permits the recognition of the present, the past and the future.

The sense of time together with the sense of space result in the sense of motion. The linear motion has been central to physics studies since the beginning of physics, that is, for at least two and half millennia. The spinning motion has been given less attention, coming into fundamental physics as late as about one hundred years ago.[d] That also may relate to physicist's instinct of survival. Suppose a physicist sees a highly-polished-spinning steel ball at a distance. S/he would not duck for the instinct of survival would signal no discernable danger to the physicist. That would be for the ball wouldn't be growing in size. While the sense of linear motion is an obvious attribute of the human mind, the sense of spinning motion is less pronounced. If the steel ball is replaced with a spinning orange, the physicist would perceive the spin of the orange as linear (orbital) motions of skin prickles. S/he might conclude that the orange is spinning but s/he wouldn't sense the spin itself. The culprit here could be the biological evolution that didn't equip the physicist with the sense of spinning motion as such a motion presents no discernable threat to survival.

Another physicist's sense of crucial importance to doing physics is *the sense of material energy*. Material energy, denoted "U," will be introduced and discussed later, in section 22. The reason for getting ahead of myself and referring to energy U at this point is that I want to draw reader's attention to the sense of energy with reference to physicist's mind. The idea of the existence of *material* energy U that fills the entire universe is that it is the same energy that the proton, the electron, and the photon consist of. Energy U can exist in three forms: in the form of the cores of the proton and the electron, denoted U_{solid}; in the form of the photon, denoted U_{photon}; and in the form of a fluid-like substance, denoted U_{fluid}

[d] That the concept of spin came to physics as late as the early 20th century, isn't really true. The spinning motion came into astronomy, for instance, in the 16th century with Copernicus explaining the motion of the earth. Later, Newton examined the motion of a spinning top when discussing his first law of motion.

that fills the entire universe between the cores of the electrons a and the protons. The point that I am trying to make here is this: The physicist can easily sense energy U_{solid} using the sense of touch. When s/he touches something, that something will be the protons, the electrons, and the neutrons, perhaps with some U_{solid}-free space between the hand of the physicist and the particles. The physicist also is able to sense photon's energy U_{photon}, but in the range of frequency higher than the microwave frequency only. S/he can sense it via the sense of touch (when the photon collides with her/his skin), or via the sense of sight/heat. The physicist has no sense of the photons with frequencies in the radio-microwave range. Those photons appear to present no discernable threat to survival. Apparently, the physicist has no sense of material energy U_{fluid} as energy U_{fluid} that we are all submerged-in presents no discernable threat to survival either.

Thus, the human mind has no sense of energy U_{fluid}. Accounting for the lack of that sense will allow for simple explanations of a number of well-known mysteries in physics, for instance, coronal heating problem, the electron-nucleus interaction, the direction of the force of gravity, and the double-slit experiment.

It needs to be said that the experimental physics did produce effective substitutes for the lack of physicist's sense of spin, the energy of photon (U_{photon}) in the radio-microwave range, and the energy U_{fluid}, as will be discussed later in the book.

6 Controversies

Some new and re-invented interpretations of several basic physics concepts critical to this discussion are pointed out in this section. They are pointed out up-the-front because they are expected to be most controversial. That will provide some guidance to the reader on where to look for a fatal flow in the suggested representations of the fundamental physics and the universe.

Nature doesn't own any attributes of the human mind. She is as simple as binary "0 or 1."

The laws of nature include the second law of thermodynamics, the action-reaction law, and the material energy-conservation law.

The range of gravity (the range of gravitational interaction) is finite. It is equivalent to the amount of gravitational potential and has a dimension of [L^3].

The speed of gravity (the speed of gravitational interaction) is infinite, while gravitational waves move with the speed of light.

Spacetime is an abstract-mathematical notion. It exists in the physicist's mind only. It doesn't affect, and it cannot be affected by anything.

Space and time that exist in the human mind should be taken as physically real. Nature, however, doesn't realize space and time.

A fictitious force is an abstract-mathematical notion. It exists in the physicist's mind only. It cannot affect anything.

The real (non-fictitious) force of inertia balances any force that causes the motion of matter such that the action-reaction law is obeyed. Unbalanced forces don't exist. The real force of inertia is generated by the gravitational interaction of an object in motion with distant masses (Mach's general idea).

Gravitational and inertial masses are the same thing. There are two distinct physical laws of gravitation formulated by Newton. Both those laws incorporate the same mass.

Gravitational-potential field is immaterial, perfectly isotropic and homogenous. It comprises no numbers. It also represents a physically-real absolute frame of reference.

The direction of the force of gravity, or its component, must be in the direction of the motion of gravitationally interacting object.

Material energy (dimension [M]) exists as a physically real substance and fills the entire universe. Material energy and matter (mass) are the same thing.

The photon comprises material energy (matter).

The standard concept of energy (dimension [$M \cdot L^2 \cdot T^{-2}$]) is the ability to do an amount of work. It applies to the abstract concepts of kinetic and potential energies.

Heat is material energy while the *measurement of temperature* is the measurement of the density of material energy.

Concerning the issue of mathematics in physics, which is a king of all controversies, it is of interest to note Lee Smolin's belief: "Mathematics is a great tool, but the ultimate governing language of science is language."[14] Many physicists share Smolin's belief. The follows also from FLH (section 2).

CHAPTER II – GRAVITATION

7 Some aspects of the theories of gravitation

The descriptions of gravitational effects are normally based on the suppositions that the range of gravitational interaction (the "range of gravity") is infinite and the speed of gravitational interaction (the "speed of gravity") is finite. Also, gravitational waves are believed to comprise ripples in immaterial spacetime, which if detected would imply spacetime to be a physical reality. Those three suppositions present virtually unchallenged cornerstones of the current theories of gravitation. It seems unlikely, however, that the correctness of any of those suppositions could be confirmed, experimentally or otherwise, in the foreseeable future. Alternative suppositions—that is, the range of gravity is finite, the speed of gravitational interaction is infinite, and gravitational waves are material—may lead to more justifiable interpretations of observed gravitational effects. Those alternative suppositions are central components of the theory to be discussed.

The well-established approach to the formulation of a theory of gravitation, which was originally conceived by Newton[15] and later followed by Einstein in the development of GR[16], has been highly successful in arriving at very accurate descriptions of gravitational interactions. That approach comprises developing mathematical models designed to describe gravitational effects. The models are designed to fit the results of experiments, observations, as well as some unconfirmed suppositions. Newton's approach is generally followed in modern theories of gravitation, including the proposed modifications to both GR and Newton's theory of gravitation.

Numerous theories of gravitation have been proposed since the time of Newton. After GR was published in 1916, most of those proposals focused on modifications to the GR field equation (e.g., T. Kaluza[17], O. Klein[18], C. Brans and R.H Dicke[19], J. Moffat[20]). In fewer cases, modifications to Newton's theory of gravitation were proposed (e.g., M. Milgrom[21], J. D. Bekenstein[22]). Key objectives of those modifications were to account for the inertia effect and/or to incorporate dark matter and dark energy, to explain the large-scale structure of the universe, and to model the evolution of the universe using variable with time gravitational constant and the speed of light.

The premise of a finite range of gravity suggested herein is not new. It was proposed on few occasions in the past (e.g., by P. G. O. Freund et al.[23], D. G. Boulware and S. Deser[24], S. V. Babak and L. P. Grischuk[25], A. D. Allen[26]). In those proposals, a finite range of gravitational interaction was assumed to make modifications to the GR field equation, which were designed to address specific aspects of the standard cosmological model, such as the black hole horizon, the large scale of the universe, or the galaxy rotation curves. In the theory proposed herein, a finite range of gravity is concluded from the consideration of a gravity paradox (section 1) rather than assumed.

New theories of gravitation, including modifications to GR, are probably published on at least a monthly basis. Some of them put forward new ideas concerning gravity that are of special interest to this discussion. For instance, the search for a connection between gravitation and quantum entanglement (S. Bose *et al.*[27], C. Marletto and V. Vedral,[28] E. Verlinde[29], S. M. Carroll and R. N. Remmen[30]) is one of those ideas. It will come out, in an entirely different and significantly simpler context, as a conclusion of the proposed theory.

The principle of equivalence, which is central to any theory of gravitation, will be shown to be a simple implication of the basic hypotheses underlying the EE theory. While normally believed to be exact, it will be argued that the principle of equivalence cannot be exact owing to a finite range of gravity. It will be suggested that the exactness of the principle of equivalence might fail at a level of accuracy of approximately 10^{-15}, or less. The accuracy of the principle of equivalence was experimentally determined in a torsion-balance test to a 10^{-13} level by T. A. Wagner *et al.*[31]. That determination was recently revised to 10^{-14} from a microsatellite experiment as reported by P. Touboul *et al.*[32]. The experimental testing of (weak) equivalence principle has been carried out with the primary purpose of addressing the issue of gravitational versus inertial mass. In the theory proposed here, that isn't an issue as inertial mass and gravitational mass are suggested to be one and the same thing.

Another key aspect of the proposed EE theory is gravitational potential. Gravitational-potential field that will be introduced and discussed in the next section cannot be discrete or quantitative. That is in agreement with the experimental results that failed to reveal the existence of graviton (J. Q. Quach[33], R. A. Norte[34]).

The most important recent observation, which is central to this discussion, was the detection of gravitational waves in the famous LIGO experiment.[35] It will be suggested later that the gravitational waves detected in LIGO experiment comprised physically-real material energy (matter) rather than abstract spacetime.

8 Finite range of gravity

Gravitational object, which will be called "g-object," is defined as an array of gravitationally-bound, finite-size, single-mass objects that don't directly interact with any single-mass objects located outside the array. As an example, the solar system would be a g-object if it is located sufficiently far from other stars such that no *direct* gravitational interactions between the components of the solar system—the sun, the planets, etc.—and other stars can occur. (Direct versus indirect gravitational interactions will be explained in section 16. Indirect interactions aren't important prior to that section. Direct interactions are those addressed in Newton's law of gravitation.)

Because the range of gravity is finite (section 1), the state of no-direct-gravitational interaction between a g-object and an outside single-mass object is possible: Simply, if the g-object is located too far from a single-mass object for their gravity ranges to be connected, neither the g-object as a whole nor any of its parts can directly interact with the single-mass object. Consequently, *there must exist a continuous boundary surrounding any g-object that delineates its gravity range in all directions. The delineated space represents the g-object's volumetric range of gravity (dimension [L^3]). It also represents the g-object's gravitational-potential field (G-P field) in the sense that it plays the role of the mediating agent underlying gravitational interactions*: As G-P field determines g-object's range of gravity, G-P fields of two g-objects must be connected such that g-objects can directly interact. It follows that the volume of G-P field equals the amount of gravitational potential (V) associated with a g-object. Each g-object must have a well-defined G-P field of finite size, which means that the boundary of G-P field must be fully closed. Thus, each g-object has a finite amount of gravitational potential.

The G-P field of a single-mass object labeled m_0, which is the simplest g-object, is shown in Fig. 1. The G-P field of a g-object that comprises four single-mass objects, denoted g-object m_{1234}, is illustrated in Fig. 2. The dashed-line circles in Fig. 2 illustrate the boundaries of individual G-P fields of the four objects. The continuous line illustrates the boundary of G-P field of g-object m_{1234} as a whole. An array comprising more than one single-mass objects will be called a "g-object." A single-mass object will be called a "mass object" or, simply, an "object."

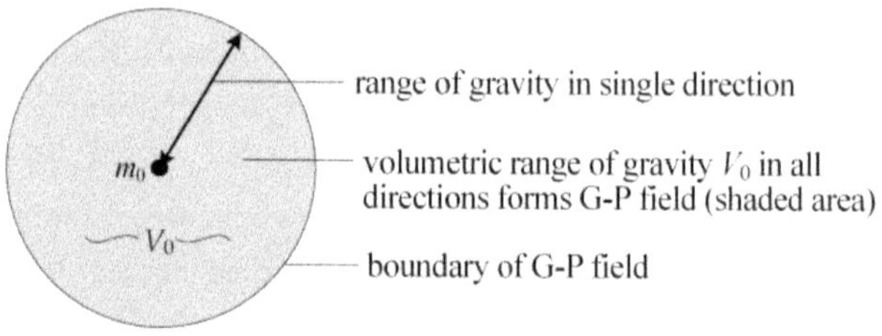

Fig. 1 G-P field of single-mass object m_0. Volume V_0 (the volumetric range of gravity) corresponds to the amount of gravitational potential equal to m_0/D.

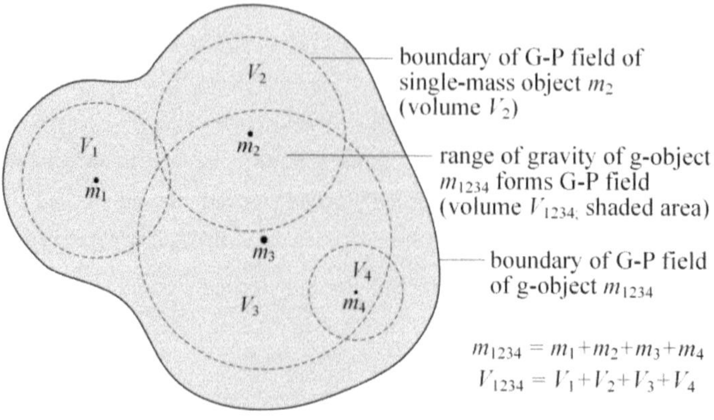

Fig. 2 G-object m_{1234} comprises four single-mass objects. G-P field of g-object m_{1234} is outlined by continuous line that delineates g-object's range of gravity.

The second and the third hypotheses stated in section 1 lead to the following postulate,

$$m = DV, \tag{1}$$

where m is the intrinsic mass-energy (mass) of a g-object, D is a proportionality constant, and V is the volumetric range of gravity, which corresponds to the amount of gravitational potential.

The key idea underlying postulate (1) is that it isn't the range of gravity of a g-object in a single direction but the "entire sum" of single-direction ranges that add up to the amount of gravitational potential V, which is proportional to the contents of the matter (mass) of a g-object in agreement with the third basic hypothesis (section 1). *The "entire sum" reflects the fact that nature operates in the directions of space. Consequently, the entire gravitational potential of a g-object comprises the "entire sum" of directional gravitational potentials (directional ranges of gravity).*

Constant D in Eq. (1) has a *dimension* of mass density $[M \cdot L^{-3}]$, and it represents the average mass density in G-P field. Eq. (1) presents the simplest, directly proportional relation between the mass of a g-object and its gravitational potential. It can be shown that the *directly* proportional relation between mass of a g-object and its gravitational potential V is the only physically satisfactory scenario. Postulate (1) states that the amount of gravitational potential (V) is a real physical property of a g-object for that property is invariant with respect to the distribution and motion of the single-mass objects that form a g-object.

G-P field that determines the gravitational potential of g-object is immaterial as it comprises no mass nor energy (i.e., no matter). It comprises no numbers (values). G-P field comprises immaterial information concerning the presence of gravitational potential. It states that either "gravitational potential is here" or that "there is no gravitational potential here," which are purely nonquantitative statements. That information exists as a physical reality, which will be verified in section 17. There is no other information associated with G-P field. G-P field is exactly the same at each point of the field in space and in time. It is perfectly isotropic and homogeneous. (It is of interest to note that according to Newton, gravitational field could be either material or immaterial.[36])

9 Physical law for direct gravitational interaction

Newton's law of gravitation is normally stated in the form of the gravity-force equation,

$$F_g^\delta = G\frac{m_1 m_2}{d^2}. \tag{2}$$

Superscript δ in Eq. (2) is meant to emphasize that Newton's force of gravity, F_g^δ, is a *direct force of gravity* rather than the true force of gravity, F_g, the importance of which will be discussed in section 14. A modified Eq. (2) will be derived in this section from the consideration of the three hypotheses underlying the EE theory (section 1). That derivation isn't intended to rediscover Newton's law of gravitation (2). The objective of presenting that derivation is to demonstrate that Newton's law (2), the suitability of which under typical conditions has been confirmed to a very high degree of accuracy, is in agreement with the implications of the three fundamental hypotheses. Said another way, the objective is to check the reasonability of the hypotheses stated in section 1 using the well-confirmed suitability of Newton's law of gravitation.

It is impossible to conclude from Newton's law that the mass of one of two gravitationally interacting objects contributes to the strength of gravitational interaction—that is, to the strength of the direct force of gravity acting between the objects—more than the mass of the other object. Neither such a conclusion can be drawn from observations or experiments. That is consistent with FLH for nature is unable to measure and thus know the amount of mass as she is unable to make measurements: The execution of an FLP (law II, in this case) is nonquantitative, as pointed out in section 2. It is therefore suggested that *the contributions of two objects to the direct force of gravity acting between them should be taken as equal in deriving a physical law for the gravitational interaction.* That suggestion follows the FLH corollary proposed in section 2 because nature, in the statement of the physical law to be derived, will be unable to distinguish between the amounts of masses of the objects. That notion will be called "inference (a)."

The concept of the agent that mediates gravitational interaction referred to in the statement of law II (section 3), can be explained with reference to a direct gravitational interaction using Figs. 3

and 4. It is G-P field, which is immaterial while physically real, that plays the role of the agent. In Fig. 3, G-P fields of m_1 and m_2 aren't connected, whence there cannot be direct gravitational interaction between those two objects according to the statement of law II. In Fig. 4, G-P fields of the two objects are connected, and the objects are subject to direct gravitational interaction.

Objects m_1 and m_2 in Fig. 3 are separated by distance d_L. That is the maximum distance at which two objects can directly interact for the two G-P fields are adjacent but not connected to each other.

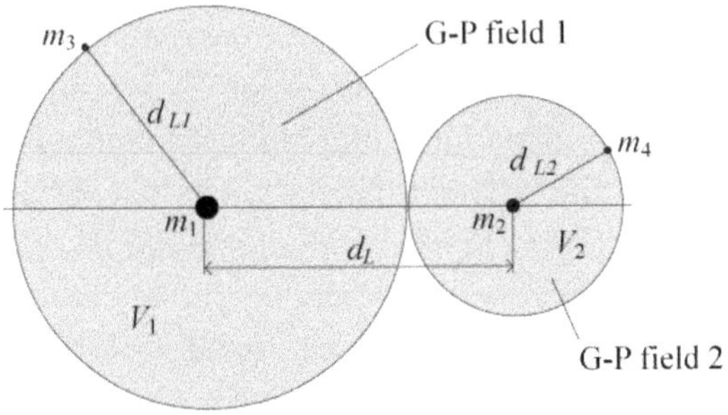

Fig. 3 Direct interaction between objects m_1 and m_2 starts or ends at distance d_L. (Ignore objects m_3 and m_4 until section 20.)

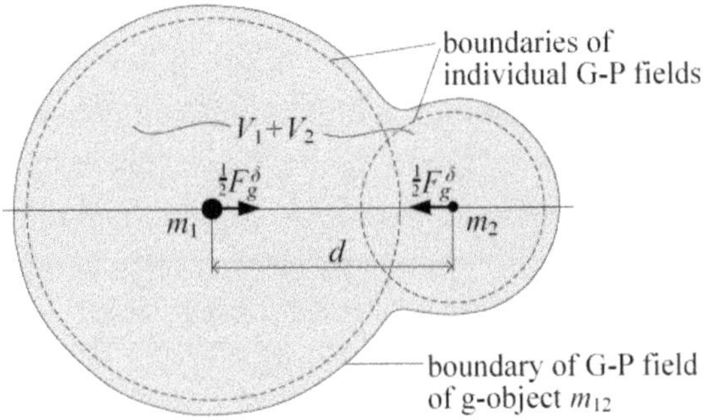

Fig. 4 At $d < d_L$ (refer to Fig. 3) objects m_1 and m_2 are subject to direct gravitational interaction.

At distance d_L, the direct force of gravity between the two objects is nil. It can have a non-zero value only if the individual G-P fields of the objects are connected (Fig. 4), that is, if $d < d_L$.

Let the mass of one or both objects shown in Fig. 4 increase. Then, the strength of the direct force of gravity F_g^δ increases in a proportion to the increase in the mass of one of the two objects, or both, which is a well-known fact of experience. That the strength of F_g^δ increases with an increase of the mass of either one or both objects will be called “inference (b)," which is suggested from the consideration of experiments and observations.

Inference (b) can also be stated in another way: Let the degree of equilibrium of the system of two objects illustrated in Fig. 4 increase. That is an equivalent statement of inference (b) for an increase in the degree of equilibrium of the system in an instant of time is equivalent to an increase in the mass of one of the two objects, or both. What that means is that the force required to bring a system of interacting objects toward nonequilibrium increases. Said another way, the higher is the degree of equilibrium of a system, the stronger is the force that binds the components of the system. The latter statement of inference (b) is more appropriate than the former for nature doesn't realize mass while, according to laws I and Ia, she realizes equilibrium.

Because nature executes her laws in instants of time and space (section 4), the FLH corollary requires that a physical law to be derived excludes, in its statement, the instances in time and the instances in space in which nature would be unable to execute the fundamental law of gravitation. Those two exclusions will be called “inference (c)."

According to the FLH corollary, a physical law should be such that whatever nature cannot possibly do should be prohibited by the statement of the law. Therefore, by the statement of a physical law for direct gravitational interaction, objects cannot be allowed to interact directly if they are outside their gravity range, that is, if their G-P fields are not connected. That suggestion, derived from the hypothesis of a finite range of gravity (section 1), will be called "inference (d)."

In the following, the objective is to construct a physical law for direct gravitational interaction that is consistent with postulate (1) and inferences (a) through (d).

Postulate (1) implies that four physical variables (m_1,V_1,m_2,V_2) are inherent to the direct gravitational interaction between two objects. It is thus rational to suggest that the strength of the direct force of gravity (the strength of direct gravitational interaction) can be expressed as

$$F_g^\delta = f(m_1, V_1, m_2, V_2, d, t), \tag{3}$$

where d and t describe space (distance) and time. To account for inferences (a) and (b), the four variables included in postulate (1) can be arranged into two terms, m_1V_2 and m_2V_1, with force F_g^δ being proportional to their sum,

$$F_g^\delta \propto \frac{1}{2}m_1V_2 + \frac{1}{2}m_2V_1. \tag{4}$$

The two terms represent contributions to the direct force of gravity acting between objects subject to direct gravitational interaction. According to postulate (1), those two terms always have the same value, which accounts for inference (a). From relation (4), it follows that the direct force of gravity F_g^δ is proportional to either of the two masses, or both, whence inference (b) is accounted for.

To derive a physical law for gravitational interaction, space (distance) and time must be accounted for in proportionality (4). Let each of the four variables identified as m_1, V_1, m_2, and V_2 in postulate (1) be averaged over the distance d shown in Fig. 4,

$$F_g^\delta \propto \frac{1}{2}\frac{m_1}{d}\frac{V_2}{d} + \frac{1}{2}\frac{m_2}{d}\frac{V_1}{d}. \tag{5}$$

The meaning of averaging over distance d will become clearer after the concept of material energy is introduced in section 22. In brief, the direct gravitational interaction between two objects can take place only if material energy, which is bound to the objects and contained along distance d in the form of equilibrium energy, is released in accordance with law Ia (section 3). No equilibrium energy exists outside distance d. That restricts direct gravitational interaction between two objects to within distance d and satisfies inference (c): Law II can only be executed over the part of the instance of space from which equilibrium energy can be released. That is consistent with the FLH corollary.

Let each of those four variables be averaged also over time period t_L, which is the longest time over which two mass objects can quantitatively interact, along the longest possible distance d equal to d_L (see Fig. 3). That leads to

$$F_g^\delta \propto \frac{1}{2}\frac{m_1}{t_L d}\frac{V_2}{t_L d} + \frac{1}{2}\frac{m_2}{t_L d}\frac{V_1}{t_L d}. \tag{6}$$

Thus, time period t_L restricts time instants to those in which direct, quantitative gravitational interaction between two objects can take place. In other words, there cannot be quantitative gravitational interaction in any time instant outside the period of time t_L. Including time t_L in proportionality (6) satisfies inference (c).

Averaging of the four variables over distance and time means that the execution of the physical law for direct gravitational interactions is restricted in space and time. It follows that a more appropriate form of Eq. (3) would be obtained if each of the four variables is reduced as follows,

$$F_g^\delta = f\left(\frac{m_1}{t_L d}, \frac{V_2}{t_L d}, \frac{m_2}{t_L d}, \frac{V_1}{t_L d}\right). \tag{7}$$

To satisfy inference (d), let a function $K = K(d_L, d)$ be inserted into proportionality (6) as follows,

$$F_g^\delta \propto \frac{1}{2}\frac{m_1}{t_L d}\frac{V_2}{t_L d}K + \frac{1}{2}\frac{m_2}{t_L d}\frac{V_1}{t_L d}K. \tag{8}$$

Function K must satisfy the following conditions: $K = 0$ for $d = d_L$ and $K > 0$ for $d < d_L$. For all $d > d_L$, force F_g^δ has to be nil and proportionality (8) doesn't apply. That accounts for inference (d) since, according to postulate (1), it is impossible for a direct force of gravity to have a non-zero value at any distance $d > d_L$.

Postulate (1) and proportionality (8) present the physical law for direct gravitational interactions. They can be combined to get

$$F_g^\delta = \frac{1}{t_L^2 D}\left(\frac{1}{2}\frac{m_1 m_2}{d^2}K + \frac{1}{2}\frac{m_2 m_1}{d^2}K\right), \tag{9}$$

where $1/(t_L^2 D)$ is a constant. For the purpose of this discussion, the simplest, linear form of function K will be assumed,

$$K(d_L, d) = k\,\frac{d_L - d}{d_L}, \tag{10}$$

where k is a dimensionless constant. More complex forms of function K could lead to more accurate predictions of the law of gravitation (9). Yet, in most applications law (9) is sufficiently accurate for it has to be at least as accurate as Newton's law of gravitation. Assumption (10) is introduced to arrive later at some conclusions that will not be sensitive to the accuracy of law (9). For $K = 1$, Eq. (9) becomes,

$$F_g^\delta = G\frac{m_1 m_2}{d^2} \text{ where } G = \frac{1}{t_L^2 D}, \tag{11}$$

which is the same as Newton's law of gravitation (2). From Newton's second law of motion given by $F_m = ma$, and Eq. (11) assuming $F_g^\delta = F_m$, acceleration of a free-falling object, measured in relation to the object toward which the object is falling, is:

$$a_i = [(1/t_L^2 D)(m_j/d^2)]K. \tag{12}$$

There seems to be no indication that Newton's original proposal $F_g^\delta = F_m$ is justifiable. Einstein believed that equality to be correct as he proposed, in GR, that gravitational and inertial masses are numerically equal. In section 18, it will be suggested that $F_g^\delta = F_m$ is a firm consequence and a central feature of the proposed EE theory, while gravitational and inertial masses are the same thing.

From Eqs. (10) and (12) for $k = 1$, and for densities D at 3.0×10^{-19} and 4.0×10^{-28} kg·m^3, the values of function K for the sun-Pluto interaction are 0.9994927 and 0.9999994, respectively.[e] For other planets, the K values are closer to unity. It follows that approximation $K \approx 1$ applies well to the sun-planet interactions. That is consistent with the high accuracy of Newton's law (2), which was derived in part from the observations of sun-planet interactions. As in the earth-based laboratories the value of function K is close to unity, Eq. (9) in combination with Newton's $F_g^\delta = F_m$ is expected to be highly accurate as well. On a very small scale, recent experimental results showed that Eq. (9) with $K = 1$ is highly accurate to a separation distance of 52 μm (Lee *at al.*[37]).

The primary purpose of presenting the derivation of Eq. (9) has been to show that inference (a), which is a key to that derivation,

---

[e] Those values of average mass density D in G-P field are suggested to represent the upper and lower boundaries of the actual D value (section 63).

leads to a very good agreement with the results of all experiments and observations against which Newton's law of gravitation has been tested. That supports FLH's implication that the executions of laws of nature are nonquantitative.

As suggested, time period t_L is the time of free fall over the longest possible direct interaction distance d_L. According to the proposed law of gravitation (9), the value of t_L is independent of the masses of interacting objects. Assuming simplification (10), approximate calculations show that the values of t_L calculated directly from gravitational constant G (denoted t_{L_a}) and from using Eqs. (10) and (12) (denoted t_{L_b}) are related as follows: $t_{L_a} = 0.175 t_{L_b}$. Hence, the value of constant k in Eq. (10) would be 32.6,

$$K(d_L, d) = 32.6 \frac{d_L - d}{d_L}, \tag{13}$$

in which case $t_{L_a} = t_{L_b}$.

The structure of constant $1/(t_L^2 D)$ in the law of gravitation (9) appears to be analogous to that of Newton's constant G expressed in Planck units,

$$\text{Planck units: } G = \frac{l_P^3}{m_P t_P^2} = \frac{V_P}{m_P t_P^2} \qquad \text{Eq. (9): } G = \frac{1}{D t_L^2} = \frac{V}{m t_L^2},$$

where l_P, m_P and t_P are Planck units, V_P is Planck volume, m is the mass of a g-object, V is the amount of g-object's gravitational potential, and t_L is the longest period of time over which direct gravitational interaction can take place.

CHAPTER III – IMPLICATIONS OF THREE BASIC HYPOTHESES

10 Action-at-a-distance mode of interaction

A frequently used argument against Newton's law of gravitation is the alleged action-at-a-distance mode of interaction. The premise of gravitational field introduced in section 8 overturns that allegation. An action-at-a-distance interaction could be alleged if two objects are separated by distance $d \geq d_L$ (Fig. 3), and interact directly. In that case, however, no direct gravitational interaction can take place as G-P fields of the two objects are separated, that is, the mediating agent doesn't connect the objects. At distances $d < d_L$, the objects directly interact as their G-P fields are connected (Fig. 4), whence the supposition of an action-at-a-distance mode of interaction isn't relevant. Note that no direct gravitational interaction can occur in the absence of a common G-P field connecting two objects along the straight line that extends between the objects. That straight-line condition has to be satisfied for dehumanized nature operates in instants (directions) of space, as discussed in section 4.

11 The speed of gravity

As nature doesn't realize space, she is unable to realize distances between gravitationally interacting objects. As a result, nature is expected to execute quantitatively law II (section 3) such that two objects interact regardless of the distance that separates them, provided that the objects are connected by G-P field and a conflict exists. Furthermore, nature operates in instants of time (section 4). To afford gravitational (or any other) interaction in an instant of time over a distance, that interaction has to be instantaneous:

$$speed\ of\ gravitational\ interaction = \infty. \tag{14}$$

The infinite speed of gravity will also be concluded in section 11 from another consideration. Keep in mind that the speed of gravity means the speed of gravitational interaction. It shouldn't be confused with the speed of the gravitational wave.

12 Gravitational-potential-free regions in the universe

A gravitational-potential-free region (G-P-FR) can exist where all mass objects in the universe are located too far from that region for their combined G-P field to extend into it. That G-P-FRs have to exist in a spatially-infinite universe will be discussed shortly. Assuming isotropy and homogeneity of the universe on a large scale, which means that the cosmological principle holds, means that the distribution of G-P-FRs in the universe is isotropic and homogeneous on a large scale.

Let star S be in motion through the G-P field as illustrated in Fig 5. As seen from the star's location, the G-P field extends to the boundary of a G-P-FR in any direction. The arcs in Fig. 5 form a discontinuous gravitational-potential-free horizon (G-P-FH) of the star.[f] As the star moves, the G-P field doesn't move or deform

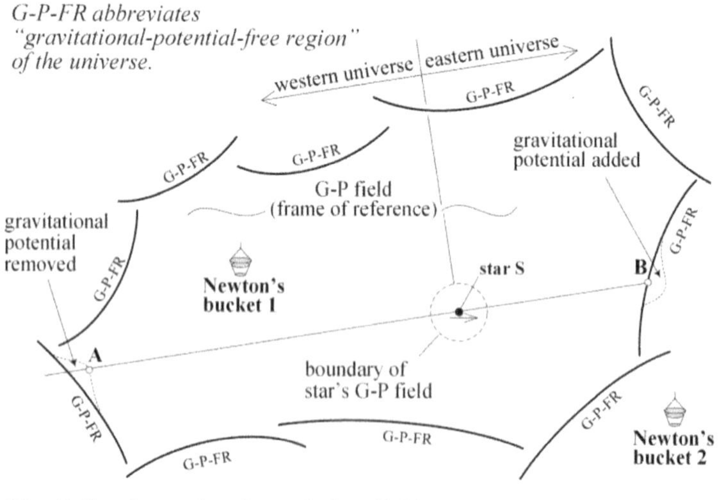

Fig. 5 Star in motion through G-P field. Space confined by the arcs forms a "supercluster." The arcs form a discontinuous gravitational-potential-free horizon. (Ignore Newton's buckets until section 17.)

[f] To visualize the concept of G-P-FH, imagine standing in the middle of a forest and shooting a rifle in all directions. While the tree barrier is discontinuous, a bullet will hit a tree in any direction it is fired. As a bullet cannot hit anything behind a tree, a g-object cannot interact with any g-objects "hidden" behind object's G-P-FH. That is for nature executes her laws in directions of space (section 4). For two objects to interact, a mediating agent, which is G-P field introduced in section 7, must connect the objects. That is, it must be continuous along the direction of the connecting straight line. If G-P-FRs exist indeed, gravity forces acting on an object cannot be infinite.

for *the information about gravitational potential remains exactly the same at each point of the field*, as discussed in section 8. That information is immaterial and nonquantitative, and there is no any other information or property associated with the G-P field. In effect, the G-P field of a star is stationary in the frame of the star. Also, the G-P field of any material object is stationary in the frame of the object.

The amount of gravitational potential *V* of an object has to be conserved according to postulate (1) as the amount of mass has to be conserved. It follows that as star S is moving, a part of the G-P field (part of the information on the existence of gravitational potential) is removed from somewhere behind the star and added somewhere in front of it as indicated at locations A and B (Fig. 5), respectively. The remainder of the G-P field remains absolutely unchanged and motionless in relation to star S, and to any other object in the G-P field. It follows that G-P-FRs have to exist in the universe such that postulate (1) is always satisfied. Without those regions, average mass density *D* would increase in the eastern part of the universe while decreasing in the western part (Fig. 5), which would violate postulate (1) for *D* is a constant. Because G-P-FRs are distributed homogeneously on a large scale throughout the universe, any mass object in the G-P field must be surrounded by its own discontinuous G-P-FH.

Without G-P-FRs any object in the spatially-infinite universe would be subject to an interaction with an infinite amount of mass in each direction. An infinite amount of mass interacting with an object also means infinite strength of the force of gravity acting on the object in each direction. Such a scenario has never been observed, which implies that it isn't the case. The suggestion of a finite range of gravity rid of the rather uncomfortable problem with having forces of infinite strength acting on each object. That was the problem that Einstein referred to in his critique of Newtonian gravitation, in which both *the universe and the range of gravity were assumed to be infinite*.[38] Such forces would cancel out with respect to a single object. However, with respect to two interacting objects, a no-relative-motion condition would exist. It is therefore suggested that the forces of gravity which can potentially act on an object are limited to those resulting from the interactions of the object with masses contained inside the object's G-P-FH, rather than infinite masses.

As pointed out above, gravitational potential V has to be added in front of a moving star and simultaneously removed behind the star for gravitational potential V has to be conserved in agreement with postulate (1). Hence, *the speed of gravitational interaction (the speed of gravity) has to be infinite for gravitational potential V has to be conserved at each instant of time*.

It is convenient to introduce a special concept of a supercluster. It is different from the common concept, except that both mean relatively large parts of the universe. The G-P field and the masses surrounded by the arcs in Fig. 5 will be called a supercluster. A "supercluster" is defined with respect to a single-mass object. Any mass object located in the G-P field of the universe has its own supercluster. Only the stars that have the potential to interact with star S are parts of the supercluster of star S. For the potential to exist, two objects (star S and another star) must be connected by a mediating agent, which is the G-P field. As nature operates in instants (directions) of space, a continuous G-P field must connect two objects along the straight line extending between the objects for those objects to interact or have a potential to interact.

13 Absolute frame of reference

The G-P field introduced in section 8 presents an absolute frame of reference with respect to gravitational interactions. That is for the G-P field between points A and B in Fig. 5 remains absolutely unchanged and motionless as the star or any other object located in the field moves in relation to the frame of the field. As discussed before, the G-P field comprises nonquantitative information, which reads: "Gravitational potential is here." That information remains exactly the same at each point of the G-P field regardless of the mass and the motion of any object. In particular, it remains exactly the same inside and outside the G-P field of a star (the dashed circle in Fig. 5). In section 18, it will be suggested that any single-mass object located in the G-P field of a g-object (e.g., in the G-P field of a galaxy) has to be in motion in the frame of any other object located in the field.

14 The force of antigravity

Consider a galaxy, spiral or elliptical. There are forces of gravity pulling stars in the galaxy toward the galaxy's center. Let those forces inside the galaxy be switched off. The mass of the galaxy is in a state of nonequilibrium as it is contained primarily in the galaxy's bulge. Under that scenario, what would one expect to see as the result of the execution of FLPs? In particular, what motions of the stars would one expect to observe? The answer seems to be clear-cut if one accounts for the execution of the SLT: The galaxy stars would be moving away from each other, pushed by a force of gravitational-thermodynamic origin, which I will call *the force of antigravity*, F_n. The stars would be moving away from the galaxy center as nature eradicates nonequilibrium by executing the SLT, which must include the eradication of nonequilibrium in mass distribution. Let the pulling-toward-the-galaxy-center forces be switched back on. The *true forces of gravity*, F_g, and the forces of antigravity, F_n, would be pulling the stars toward the galaxy center again, as is known from observations. Those two forces are expected to act along the line of star's motion such that their resultant is balanced by the force of inertia. (The force of inertia will be discussed in section 17.) Of course, there will be other forces acting on the two stars. Regardless, it follows that one cannot detect and measure the true force of gravity, F_g, and the antigravity force, F_n, in separation. One can detect and measure the resultant of those forces only. In section 9, that resultant was named the direct force of gravity F_g^{δ}.

It is rational to suppose that the force of antigravity is mediated by the same G-P field as the true force of gravity. (Two kinds of G-P fields would be untenable from the perspective of FLH as that would require nature to make decisions.) The true force of gravity and the force of antigravity are both generated by the execution of the SLT. *Both forces drive matter toward an equilibrium (toward a no-motion) state consistent with law I.* For two gravitationally interacting objects, a no-motion state would be attained when the objects collide and stop interacting in an observable way. But, a no-motion state would also be attained if the matter becomes distributed homogeneously throughout the universe such that the objects stop interacting and the antigravity forces become nil.

Suppose that two interacting objects illustrated in Fig. 4 have been moved away from each other to the point that their G-P fields separate (Fig. 3). Before separation, there was a conflict between the true force of gravity and the force of antigravity. That conflict was necessary to generate a quantitative (observable) interaction between the objects, consistent with law II (section 3). After the separation of G-P fields, the true force of gravity became nil as the conflict disappeared. The antigravity force became nil as well.

If the force of antigravity was stronger than the force of gravity, a universe couldn't exist as all matter would be smeared out to a perfectly smooth state. There would be no local inhomogeneities. There would be nothing to observe and nobody to observe. That appears to be the key to the appreciation of the existence of any universe.

The suggestion of the existence of the force of antigravity could be seen far-reaching. Its existence, however, appears to be rational for it is well-known that the execution of the SLT applies to energy distribution, for instance, the distribution of heat energy, whence it is expected to apply to mass distribution as well.

15 The strength of the true force of gravity

The ratio of the strengths of the electric and gravity forces acting between the interacting proton and electron is

$$\frac{F_e}{F_g^\delta} = 2.25 \times 10^{39}, \qquad (15)$$

where F_e is the electric force. FLH requires that FLPs are executed in a nonquantitative way (section 2). It follows that in the absence of conflicts that would affect gravitational and electric interactions differently, the quantitative effects of the executions of laws II and III should be the same. That is, it should be that

$$\frac{F_e}{F_g^\delta} = 1. \qquad (16)$$

The ratio (15) versus ratio (16) discrepancy implies that there have to be one or more quantitatively different conflicts associated with the execution of gravitational and electric FLPs. In sections 32 and

33, it will be suggested that, in spite of the well-known similarities between the gravity and the electric force equations, those forces are generated in response to different nonequilibrium conditions such that the conflicts are quantitatively different.

It follows that a force has no magnitude (strength) per se, which is consistent with FLH: As nature laws are nonquantitative, the strength of a force must be an emergent property of an interaction. It is determined by the strength of the conflicts with the execution of a nature's law. It follows that the frequently-made statement, "the force of gravity is incredibly weak in comparison with "the electric force" isn't really pertinent.

16 Direct versus indirect gravitational interaction

Fig. 6 illustrates direct gravitational interaction between two mass objects, m_0 and m_1. The strength of the direct force of gravity acting between the objects can be determined from the physical law of gravitation (9). That law, however, should not be used to describe any aspects of *indirect* gravitational interactions. To explain that caveat, let an object m_1 be broken into four objects that are spread out as illustrated in Fig. 7. The G-P fields of those objects are no longer connected to that of object m_0, whence none of the four objects can *directly* interact with m_0.

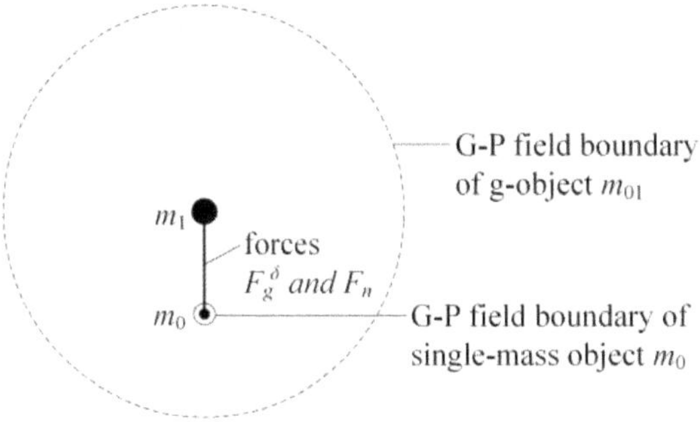

Fig. 6 G-object m_{01}. Object m_0 directly interacts with object m_1 as their G-P fields are connected and a conflict between gravity and antigravity forces is inflicted. (G-P field of m_1 isn't shown.)

However, all five objects shown in Fig. 7 are connected by their common G-P field outlined by a dotted line. Therefore, each of those objects has a potential to interact with object m_0 according

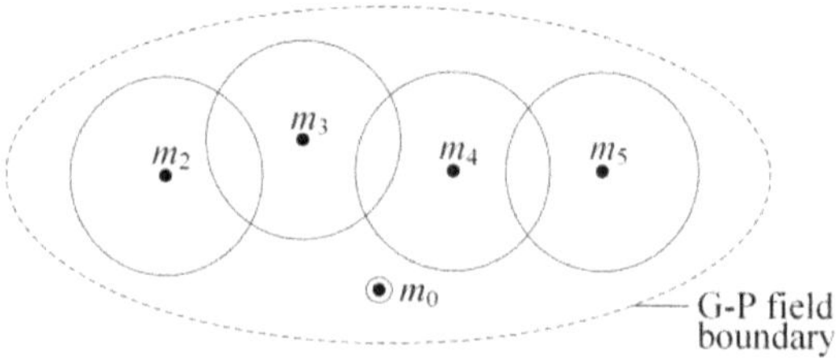

Fig. 7 G-object m_{02345}. Object m_0 doesn't directly interact with any of the four other objects as no conflicts exist. Applying a force to object m_0 would inflict a conflict. Object m_0 would then interact indirectly with the four other objects as well as with trillions of stars located within the object's G-P-FH (inside the object's supercluster)..

to law II. That would be an *indirect* gravitational interaction. To generate a quantitative interaction, inflicting a conflict is required (section 3). Let a mechanical force be applied to m_0. A conflict is inflicted and m_0 indirectly interacts with the four other objects. Let star S (Fig. 5) be replaced with the system illustrated in Fig. 7. The arcs outline now the supercluster of object m_0. Applying a force to object m_0 has to result in a quantitative interaction of the object with trillions of stars ("distant masses") located inside its G-P-FH as the stars are connected to m_0 by the supercluster's G-P field, and a conflict has been inflicted. Objects m_2, ..., m_5 are part of distant masses. As there are no antigravity forces acting between object m_0 and the trillions of stars inside its G-P-FH (the G-P field of m_0 isn't connected to a G-P field of another object), the forces of gravity involved in the interaction of object m_0 with distant masses are the true forces of gravity (F_g). It is important to realize that applying a mechanical force to m_0 under the Fig. 6 scenario also would lead to indirect interaction of object m_0 with trillions of stars. That would be what is called an inertial effect.

Considering the Fig. 7 scenario, it appears intuitively likely that the strength of indirect interaction between an object m_0 (let it be a 1,000 kg steel ball) and the four close-by objects, each having a mass, say, of the moon, is much less than the strength of indirect interaction of m_0 with trillions of stars located inside the G-P-FH of object m_0. Sciama predicted the same in quantitative terms from his theory of gravitation focused on the origin of inertia and Mach's principle.[39] He suggested that approximately 99% of the strength of the force of inertia acting on an object (such as object

m_0) is generated by the interactions with masses located farther away than 10^8 light-years from the object.

17 The force of inertia

To avoid confusion: For the purpose of this discussion, the force of inertia is understood to be the force that pushes a driver in an accelerating car into the back of the seat. In agreement with the action-reaction law, the force of inertia balances the mechanical force generated by the car's engine. A force acting on an object in uniform motion relative to the G-P field is balanced by the force of inertia owing to the action-reaction law. As nature operates in instants of time (section 4), she cannot know if an object is in accelerated or in uniform motion. As nature operates in instants of time, the balancing forces of inertia result in the universe being in perfect equilibrium in each instant of time.

Thus, the inertial effect, which results in the generation of the force of inertia, isn't an innate property of matter or spacetime. It is an emergent property of the execution of the action-reaction law. Following Mach's general idea,[40] it is suggested that the force of inertia is generated by the interaction of an object in motion with distant masses. The concept of "distant masses" will be clarified shortly.

While developing GR, Einstein believed in Mach's idea of distant masses generating local inertial effect, and suggested that GR was founded upon it.[41] Later, however, he calculated using the equation of GR that the forces exerted by distant masses on a local object were "so small that confirmation of them by laboratory experiments is not to be thought of ," and suggested that most of the inertia effect was a local property of spacetime.[42] From the perspective of the EE theory, that was an incorrect calculation. That is because GR, similar to the proposed law of gravitation (9) and Newton's law (2), is a physical law designed to describe a direct interaction between two objects subject to the direct force of gravity, and not an indirect interaction of an object with trillions of stars subject to the true forces of gravity. The correct law of gravitation for calculating the strength of interaction of a local object with distant masses will be discussed in section 18.

In current cosmology, the concept of "distant masses" appears to be vague. In this discussion, *distant masses—defined in relation to a single-mass object—are all mass objects (stars, galaxies, etc.) together with their individual G-P fields that are located outside the G-P field of a single-mass object, while inside its G-P-FH.* As an example, consider the lonely star S (Fig. 5). Its individual G-P field isn't connected to the individual G-P field of any other star. Distant masses in relation to star S are all stars located inside the supercluster of star S. That supercluster is delineated by the arcs shown in Fig. 5.

Let motionless Newton's bucket 1 hang on a rope in the frame star S (Fig. 5). Star S is at rest in the frame of the supercluster's G-P field. The rope is wound. There are trillions of stars located inside the bucket's G-P-FH, forming distant masses. Consistent with law II, a potential for quantitative interaction of the bucket with distant masses exists as the bucket and distant stars are connected by the G-P field of the supercluster. For quantitative interaction to occur, a conflict is needed. Let the rope go unwinding. The force stored in the rope is released and a conflict is inflicted. As the rope unwinds, the bucket spins. The G-P field of the supercluster represents the frame in relation to which the bucket spins. In the presence of conflict, indirect interaction of the bucket with distant masses generates a *real (nonfictitious) force of inertia* of gravitational origin, which causes the concavity of the water surface. The real force of inertia replaces the fictitious (Newtonian) force of inertia.

The bucket would be subject to the same inertial force if located close to the boundary of gravitational-potential-free space (G-P-FS)—for instance, close to point A in Fig. 5—and close to the location of star S. The fact that the majority of distant masses would be located on one side of the bucket spinning at point A would not affect the force of inertia. In a direct gravitational interaction, the force of inertia, F_i, acts in the direction opposite to a locally inflicted conflict, which arises from the true force of gravity, F_g, and the antigravity force, F_n, that is, in the direction opposite to the direct force of gravity (F_g^{δ}). That is a requirement of the action-reaction law. In an indirect interaction of an object with distant masses, the force of inertia must also be in the direction opposite to the inflicted conflict by virtue of the same law, that is, in the direction opposite to the applied force. To put it another way, the line of the force of inertia has to be the line of motion.

The force of inertia cannot exist without motion in the frame of a G-P field. That force has to be real, that is, nonfictitious for the action-reaction law has to be obeyed in real terms. The magnitude of the force of inertia must be equal to the applied force or its component in the direction of motion. It follows that the force of inertia acting on a spinning bucket is, with reference to the G-P field, location and direction independent. Its strength depends only on the magnitude of the inflicted conflict, in this case on the strength of the applied mechanical force stored in the rope.

Let Newton's bucket be spinning in a G-P-FR of the universe (Fig. 5, Newton's bucket 2). According to law II, the bucket cannot gravitationally interact with distant masses as there is no G-P field present. That is, there is no mediating agent connecting the bucket with distant masses. The water surface remains flat. Most likely, that thought experiment has never been contemplated because, in current physics, there cannot be G-P-FRs in the universe as each object interacts with all other objects in the universe, which is the Newtonian concept of gravitation summarized in law (2).

Let an object, such as an arrow or a vehicle located in the G-P field illustrated in Fig. 5 accelerate. That inflicts a conflict and the object is subject to a force of inertia that can be detected and measured. This means that the G-P field can be detected. It follows that in spite of being immaterial, the *G-P field is expected to exist as a physically real mediating agent* ("information agent" would be a better term). The second part of the proof that the G-P field is physically real is derived from a closely related experiment. Let a mechanical force be applied to an object located in G-P-FS. The object will experience no inertia force. For instance, the water surface in Newton's bucket will remain flat as there will be no G-P field (mediating agent) connecting the bucket to distant masses.

In summary, the preceding discussion explains the meaning and the origin of the force of inertia: In response to inflicted conflict, the indirect interaction of an object with distant masses generates a real force of inertia of gravitational origin, mediated by G-P field. The field exists as a physical reality and presents the frame of reference in relation to which objects are moving. The force of inertia comprises the resultant of the true forces of gravity exerted on an object by distant masses. The proposed explanation of the inertial effect rids of the following paradox: A *fictitious* force of inertia causes a *real* effect, such as the bulging of the earth at the equator. It also eliminates the following Newton's action-reaction law violation: The action revealed in a free-fall of an object has no equal and opposite (real) reaction.

The force of inertia is therefore suggested to be of gravitational origin. As discussed in section 11, the speed of gravity is expected to be infinite. That is a secure conclusion if the force of inertia is generated by the interaction of a local object with distant masses indeed. Every car driver that feels the force of inertia immediately pushing the gas pedal proves that the speed of gravity (the speed of gravitational interaction) is instantaneous.

A man falling off a roof doesn't feel any force of inertia as each molecule of his body is subject to near-the-same-gravitational acceleration. The force of inertia is generated as a result of the conflict inflicted by man's direct gravitational interaction with the earth. The man drops a stick and measures the distribution of stress along it. He discovers a stress difference, resulting from the tidal effect, between the lower and the upper ends of the stick. The man realizes that the stick and therefore himself are subject to direct gravitational interaction.

Deep in intergalactic space, an astronaut in an accelerating rocket does feel the inertia effect triggered by a conflict, which is the thrust of the rocket's engines. She feels it because the force of inertia is unevenly distributed over her body. If the astronaut is standing on the bulkhead, the inertia effect gradually spreads from the soles of her feet toward her scalp. The astronaut feels the same as if standing on the surface of the earth. She drops a stick and measures the stress exerted on it. There is none. The astronaut realizes that she is not subject to direct gravitational interaction, although she feels like she is standing on the surface of the earth. The falling man and the astronaut are subject to the same strength of the force of inertia, provided that their weights are the same and the rocket accelerates at g. *The difference is in the conflicting forces that generate the same effect of inertia consistent with the action-reaction law. One conflict is the direct force of gravity* F_g^δ *that acts on each molecule of the falling man's body in roughly the same way. The other is the mechanical force* F_m *acting on an* *astronaut's soles of feet, generated by the rocket's engines. There appears to be no reason to assume the inertial mass (of the astronaut) and the gravitational mass (of the man) to be merely numerically equal. The inertial and gravitational masses are the same thing. Both are revealed in gravitational interaction, and both comprise the same physically real material energy U. It is just the conflicting forces to which the man and the astronaut are subject to, that come from different sources.*

An apple is falling from a tree. There are two direct forces of gravity and two forces of inertia. One inertia force is acting on the apple upwards and the other one is acting on the earth downwards. The direct gravity forces acting on the apple and the earth balance out with corresponding inertia forces, that is, with the resultant of the true forces of gravity exerted by distant masses, such that the action-reaction law is obeyed.

Let a rifle be fired deep in a G-P-FS. A spring connects the rifle and the bullet. The spring gets stretched after the rifle is fired. That means there is a force acting between the rifle and the bullet. It must be a force of gravity for there are no other known forces that could be stretching the spring. (Electric and nuclear forces are out of question.) Because it's a G-P-FS, the rifle and the bullet aren't connected to any distant masses. Antigravity forces don't exist. The bullet-rifle interaction must involve a true force of gravity. From the perspective of the rifle, it is the bullet that plays the role of "balancing" distant masses, and vice versa.

18 The other physical law of gravitation

The prior discussions suggest that another law of gravitation has been long available in the form of Newton's second law of motion,

$$F_m = ma_m = -F_i, \tag{17}$$

where F_m is a force (gravitational, electromagnetic, mechanical, etc.) applied to a local single-mass object, and a_m is the resulting acceleration of the object. F_i is the force of inertia of gravitational origin, which is the resultant of the true forces of gravity acting on an object in accelerated motion. The true forces of gravity are generated by the interaction of a single-mass object with trillions of objects located inside its G-P-FH, that is, with all mass objects forming the supercluster of a single-mass object.

Consistent with law II, force F_m inflicts a conflict that results in a quantitative (observable) interaction of a single-mass object with distant masses. Therefore, *Eq. (17) presents a physical law for the magnitude of the resultant of the true forces of gravity acting on an object accelerated by force F_m. The object is in motion in the direction opposite to that resultant, which is the same as the force of inertia, F_i.* The law for indirect gravitational interactions (17) is in agreement with FLH, which implies that an inflicted conflict results in a quantitative interaction of objects if the objects are connected by a mediating G-P field. Inflicted conflict has to be of sufficient strength to put an object in motion, for instance, to start moving a car.

Law (17) is suitable for calculating the strength of the indirect gravitational interaction of an accelerated object with distant masses, that is, for calculating the strength of the force of inertia. The amount of conflict is determined by the strength of the applied force F_m, which equals the strength of the force of inertia (plus the force of friction) acting in the opposite direction to force F_m. That is a requirement of the action-reaction law.

It follows that Newton's equality $F_g^\delta = F_m$ referred to in section 9 isn't an axiom but, rather, a conclusion of the EE theory. Simply put, the direct force of gravity F_g^δ—the one that drives a falling apple down—is the conflicting force that causes the interaction of the apple with distant masses. It can be inserted into Eq. (17) to become an applied, conflicting force F_m. *It further follows that gravitational and inertial masses are one and the same.*

To illustrate the nonlocal base of the law of gravitation (17), it can be written using postulate (1) in the form,

$$F_m = ma_m = -DV_m a_m = -\frac{m_{\text{GPFH}}}{V_{\text{GPFH}}} V_m a_m, \tag{18}$$

where V_m is the gravitational potential of object m, m_{GPFH} is the total mass inside object's G-P-FH, and V_{GPFH} is the total amount of gravitational potential contained inside the G-P-FH. Law (18) states that, for a given magnitude of conflict required to accelerate an object to a_m, the force of inertia, with magnitude equal to force

F_m, is proportional to the total mass contained inside object's G-P-FH, and inversely proportional to the gravitational potential of that mass. Consistent with nature's inability to realize space (distance), the inertia force is independent of the distribution of mass objects within G-P-FH. Since postulate (1) is a consequence of the second and the third hypotheses underlying the EE theory, the nonlocality of the inertia effect (Mach's general idea) isn't an assumption underlying the EE theory. Rather, it is the conclusion of the theory. Since about the 1890s, the primary argument against Mach's principle has been that Newton's second law of motion, which determines the magnitude of the force of inertia, doesn't include a variable or a constant that could reflect the influence of distant masses. Eq. (18) refutes that argument because the average mass density in G-P field (D) reflects the total amount of distant masses inside an accelerated object's G-P-FH, that is, inside its supercluster.

Because the G-P field is the agent that mediates gravitational interactions, each object located in the G-P field has a potential to interact gravitationally with all other objects located inside the object's G-P-FH. If just one of those objects directly interacts with another object, the motions of the two objects will result in a multitude of conflicts that will cause indirect interactions between the two objects with all other objects located within their G-P-FH, and beyond. It follows all objects located inside a G-P-FH—which means *all objects located in the entire G-P field of the universe—are expected to be subject to gravitational interactions and related motions at all times. Said another way, each single-mass object located in a G-P field (e.g., in the G-P field of a supercluster, a galaxy cluster, or the solar system) is always in motion in its G-P field frame.*

19 The two physical laws of gravitation

The laws for direct (9) and indirect (18) gravitational interactions relate to two different aspects of the action-reaction law. Those, in turn, pertain to the fact that nature doesn't realize space and time (section 2), thereby enforces her laws in instants of space and time.

The direct-interaction law of gravitation (9) refers to an instant in time. It solves a static problem by describing how the action-reaction law is satisfied in an instant of time: The forces acting on a falling apple—that is, the direct and true forces of gravity, and the antigravity force—balance out in any instant of time as $F_g^\delta = F_g - F_n = -F_i$ such that the action-reaction law is always obeyed. The indirect-interaction law of gravitation (18) refers to an instant in space. It solves a dynamic problem by describing how the action-reaction law is obeyed in an instant of space: The work done on a falling apple—that is, the work done by the direct force of gravity and the work done by the force of inertia over the same distance and in the same direction—balance out as $F_g^\delta \Delta d = F_i \Delta d$, thus obeying the action-reaction law.

20 The equivalence principle

Galileo derived the principle of equivalence from experiments. The principle states that the acceleration of an object in free-fall toward another object doesn't depend on the mass of the falling object. The acceleration is measured in the frame of the other object. Galileo's principle of equivalence can also be concluded directly from FLH for nature is unable to realize the amount of mass of a free-falling object (she owns no weight scales). That conclusion, however, would only be appropriate where the initial conditions and the conflicts that represent boundary conditions applicable to free-falls of different-mass objects are exactly the same. Under the same conditions, therefore, objects with different masses should be subject to the same accelerations, for nature cannot tell one mass from another. As far as we know, the only relevant difference between two objects in free fall toward another object is their mass: According to the experience and fundamental law of gravitation II, there is no indication that either the internal structure or chemical composition of an object in free fall might play a role in gravitational interaction.

The principle of equivalence isn't a law of nature for nature is unable to realize acceleration, which is an emergent property of a free-fall interaction, and is defined in terms of space and time. The principle of equivalence is just an emergent physical law. As such, it doesn't need to be absolute and exact.

If two objects having different masses are in free fall subject to the same initial conditions and the same conflicts—that is, the same interactions with conflicting FLPs and conflicting matter—

the same quantitative (observable) effect associated with the free falls is expected to emerge. Initial conditions applicable to a free-falling object include the initial values of its speed v and distance d. Time t isn't included, for it can be chosen arbitrarily. Consider two relatively-small-mass objects m_3 and m_4 in free-fall toward large-mass objects m_1 and m_2, respectively (see Fig. 3). Each of the four objects has a mass that is relatively small or relatively large and that is significantly different from the masses of the three other objects. Objects m_3 and m_4 start falling at the maximum possible free-fall distances, denoted d_{L1} and d_{L2}, respectively. Under the initial conditions, there are no motions. The initial conditions for objects m_3 and m_4 are the same: $v_3 = v_4 = 0$, $d = d_{L1}$, $d = d_{L2}$, in the frames of m_1 and m_2, respectively.

Quantitative effects generated by free-falling objects include the speed of the object v, distance d, acceleration a, passing time t, and the maximum free-fall time t_L. As the initial times (t_0) of the free-falls can be chosen arbitrarily for each of the objects, comparisons of generated effects v, d, a and t between two free-falling objects aren't relevant. The only relevant quantitative effect is the total free-fall time t_L, which is independent of t_0.

Using Eqs. (12) and (10) with $k = 1$, the time periods t_L can be estimated for the upper and the lower bound values of density D of 3.0×10^{-19} kg·m^3 and 1.0×10^{-28} kg·m^3, respectively.[g] Masses m_1 and m_2 are assumed to be the masses of the sun and the earth. The test masses m_3 and m_4 are assumed to be 1.0×10^0, 1.0×10^1, 1.0×10^4 and 1.0×10^6 kg. For the upper bound value of density D and all combinations of the four masses, approximate calculations show time period t_L to be about 4.0×10^7 years, with a maximum difference between the various combinations of 3.0×10^1 years. For the lower bound value of D and all combinations of the four masses, time period t_L is about 2.2×10^{12} years, with a maximum difference between the various combinations of 1.8×10^6 years. As expected, the conclusion is that for a given value of density D, times periods t_L are approximately the same for all relatively-small-mass objects in free falls. That appears to present evidence in support of FLH's notion of nonquantitative executions of FLPs for nature clearly doesn't account for the amount of mass of a free-

[g] The upper and lower bound estimates of D will be discussed in section 63.

falling object: Whether the test object has a mass of 1 kg, 1,000 kg or 1,000,000 kg makes practically no difference.

Because of the finite range of gravity, the equivalence principle cannot be absolutely exact. This can be seen from Eq. (12), which shows that gravitational acceleration depends on the maximum free-fall distance d_L, an argument of function K. Distance d_L, in turn, depends on the mass of a free-falling object. (Any other form of function K would have to include d_L.). In effect, the conflicts that two free-falling objects with different masses are subject to are expected to be slightly different.

While not strictly exact, the principle of equivalence is expected to be exceedingly accurate, particularly for objects that have very small masses in comparison with the masses of the objects toward which they are falling. Assuming the form of function K given in Eq. (10) with $k = 1$, it can be calculated using Eq. (12) that accelerations of 1 kg and 1,000 kg test masses at the earth's surface would differ by 1.8×10^{-14} m·s^{-2} if density D is assumed at 3.0×10^{-19} kg·m^{-3} (they would differ by a fraction of 2×10^{-16}). If the 1,000 kg test mass is replaced with the mass of the moon, the two accelerations would differ by 7.0×10^{-8} m·s^{-2} (by a fraction of 7×10^{-9}). If density D is assumed at 1.0×10^{-28} kg·m^{-3}, the accelerations would differ by less than 1×10^{-15} m·s^{-2}.

CHAPTER IV – MATERIAL ENERGY

The new look at gravitation presented in the prior sections was based on the consequences of FLH, the resolution of the gravity paradox together with the assumed linear relation between mass and the range of gravity (section 1). Another salient paradox in physics is the gravitational wave paradox. Gravitational waves were first predicted by Poincare[43]*, then by Einstein*[44]*, and detected for the first time in the 2016 LIGO experiment*[45] *The standard interpretation of gravitational waves is that they comprise ripples in immaterial spacetime. They are immaterial, for they consist of no protons and electrons, or any other known kind of matter. In that interpretation, gravitational waves represent a mathematical construct invented to model a physical effect. According to FLH, however, nature is unable to realize space and time, and cannot do math. Assuming nature to realize ripples in spacetime means a humanization of nature, which isn't allowed here.*

Yet, without any doubt, gravitational waves are physically real and material as they strained LIGO arms. They exist independent from the human mind as the results of LIGO experiment proved. A paradox thus arises: A motion of an immaterial (abstract) object called a gravitational wave, which can exist in the physicist's mind only, caused the straining of LIGO arms, which are material. (It is like the paradox of the fictitious force of inertia that exists in the physicist's mind only, while it causes the bulging of the earth.) A resolution of the gravitational wave paradox is proposed in this chapter and the consequences of that proposal are investigated in the remainder of the book.

21 Gravitational-potential and kinetic energies

The gravitational-potential energy (PE) and kinetic energy (KE) determine the abilities to do amounts of work that can be done using a mass object. That is probably the most accepted way to define the intuitive concept of abstract energy. According to FLH, however, KE and PE are immaterial physical entities that appear to exist in physicist's mind only. (Feynman in his physics lectures called them "mathematical" or "abstract" entities.) Nature is unable to realize either PE or KE for she is unable to realize an amount of mass, space, and time, and cannot make calculations. While PE and KE have a well-defined physical meaning, which is the ability to do a certain amount of work, both PE and KE represent mathematical constructs that aren't physically real, in the sense that they don't exist outside of the human mind.

Let a stone be dropped from the Tower of Pisa. The stone is in free fall. The standard view is that while a stone is in free fall, the PE of the stone is being converted into stone's KE such that the total energy contained in the stone-earth system is conserved. However, as KE is a calculated quantity that depends on the speed of an object while PE doesn't (it depends on a distance), the notion of the $KE \leftrightarrow PE$ conversion taking place in physical terms appears to be rather unfounded. All that has been demonstrated in support of the $PE \leftrightarrow KE$ "conversion" is that in the system of interacting objects the rate of change of KE of a free-falling object is numerically equal to the rate of change of its PE with opposite sign. It is the ability to do an amount of work that is conserved (see, e.g., Feynman[46]).

The following suggestion, which is based on the existence of the real force of inertia, solves the $PE \leftrightarrow KE$ conversion problem: The rate of change of KE corresponds to the work done by the direct force of gravity, F_g^δ, as the rate of change of PE corresponds to the work done by the force of inertia, F_i. As the strength of F_g^δ equals the strength of F_i with opposite sign according to the physical law for indirect gravitational interactions (17), the work done by force F_g^δ over a distance must be equal to the work done in the opposite direction by force F_i over the same distance and along the same direction, which is enforced by the action-reaction law. In that scenario, there is no $PE \leftrightarrow KE$ conversion.

In summary, while energies PE and KE are nonexistent abstract entities (with well-defined physical meanings), their rates of change in a system of two objects in free fall toward each other are numerically equal.

The above suggestion still isn't very useful as it doesn't explain the nature of gravitational-potential energy which, according to Misner, Thorne and Wheeler[47], cannot be observed or localized. That shortcoming will be addressed in section 27.

22 Material energy *U*

Energy as a material substance does not exist in current physics, which results in a mass ↔ energy conversion paradox: How can material mass (dimension [M]) be converted into an abstract entity that is called energy (dimension $[M \cdot L^2 \cdot T^{-2}]$), which exists only in the physicist's mind? In this section, it is argued that a rational interpretation of the results of LIGO experiment[48], chemical-reactions, and phase transitions, gets rid of that paradox. *Energy that exists as a material substance will be called material energy U. Energy U is another name for matter. The dimension of energy U is the dimension of mass [M]. Energy U released/absorbed from/by a system of interacting objects in accordance with law Ia (section 3) will be called equilibrium energy U^{eq}.*

The results of LIGO experiment suggest that there is a physical entity, herein called energy U, which exists as a physically-real-material substance independent from the human mind. That means that it isn't an abstract construct. It is the material substance that arrived in the form of waves at the LIGO site and caused the straining of the LIGO arms. The results of LIGO experiment also suggest that a system of two objects in free fall toward each other is open with respect to the release of material energy U. Energy U is suggested to be material because it interacts with material objects, just as the LIGO experiment proved. Material energy U, which is the same thing as matter, has to be conserved.

As energy U interacts with mass objects, it is expected to be pulled by gravity and bound (attached) to mass-energy objects, as illustrated in Fig. 8. Pulling matter together appears to be a rational

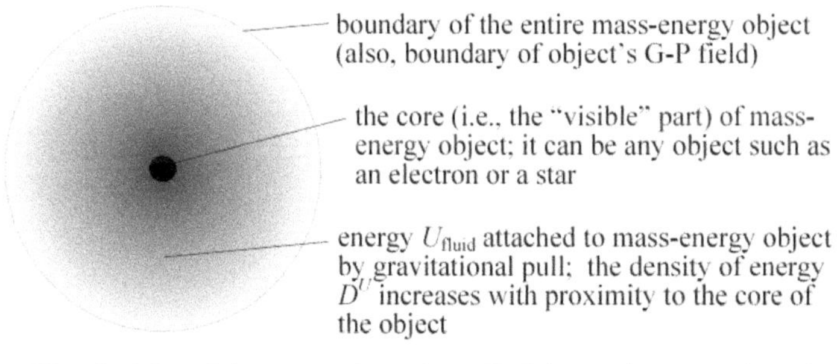

Fig. 8 Material energy U_{fluid} is pulled by and attached to the core of a single mass-energy object.

mode of gravitational interaction that can be suggested for an-object-interacting-with-energy-U scenario. It is emphasized that no other than gravitational interaction can be proposed to have affected LIGO arms for no such other interaction is known. (It certainly wasn't an electric or a nuclear interaction.)

As previously noted, the physical entity called material energy U has a dimension of mass [M] rather than the usual dimension of energy $[M \cdot L^2 \cdot T^{-2}]$. The usual dimension of energy (used in current physics) applies to the abstract concepts of energies KE and PE, which describe the abilities to do amounts of work.

That the gravitational waves detected in the LIGO experiment have to be considered material and independent from the human mind, is suggested by FLH. That suggestion is a consequence of the dehumanization of nature, which implies that a mathematical entity cannot exist as a physical reality for nature is unable to do mathematics. (Mathematics is used for and, in fact, appears to be indispensable to *describing* physical reality.) Said another way, it is trivial that a mathematical entity cannot interact with matter for such an entity cannot exist outside the human mind. Therefore, the entity that strained the LIGO arms had to be physically real and material. That appears to be an iron-clad conclusion.

The gravitational waves detected in LIGO were generated by two free-falling objects (most likely neutron stars) on the verge of colliding. The observations of binary pulsar PSR 1913+16 carried out by Taylor and Weisberg[49] suggest that gravitational waves—herein suggested to be the waves of material energy U—are released at all times during the free fall of two objects toward each other, that is, long before the eventual collision. In other words, it is expected that gravitational waves are emitted by any two mass objects in free fall toward each other.

Concerning the conservation of energy, a system of two objects in free fall toward each other isn't closed with respect to the release of material energy U. If it were, no energy would be released from such a system, whence no gravitational waves would exist as there would be no source of energy (matter) that the waves comprise.

The LIGO experiment and PSR 1913+16 observations suggest that the entire universe is filled with energy U. In current physics, the universe is filled with either quantum fields (according to QT), or spacetime (according to GR). However, as both quantum fields and spacetime appear to be mathematical constructs conceived to describe physical effects, neither exists outside the human mind, whence neither can be claimed to be a physical reality. Claiming otherwise would contravene FLH. To the contrary, the conception of the entire universe filled entirely with material energy U doesn't contravene FLH in any way.

Assume an apple (say, object m_2 illustrated in Fig. 9) is falling down from a tree toward the earth (m_1 in Fig. 9). The gravitational system to be investigated consists of the apple, the earth, and the material energy U^{eq} that is contained inside the U^{eq} region of the apple-earth system. An approximate extent of the U^{eq} region is shown in Fig. 9. Material energy U attached by gravity to the two single-mass objects, and contained within U^{eq} region, is called equilibrium energy U^{eq}. A U^{eq} region is the space between two interacting mass objects, from which material energy U can be released in accordance with law Ia.

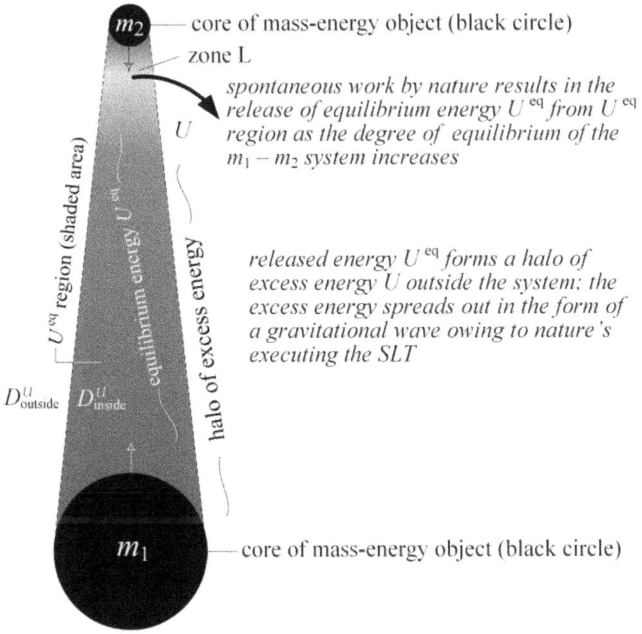

Fig. 9 Schematic of the release of equilibrium energy U^{eq} as objects m_1 and m_2 are in free fall toward each other. The average density of energy U^{eq} inside U^{eq} region is higher than average density of U in the surrounding space such that U^{eq} can be released from U^{eq} region consistent with law Ia. (Ignore zone L until section 34.)

The schematic extent of U^{eq} region illustrated in Fig. 9 reflects the implication of FLH that nature operates in instants of space (section 4), which means it operates along each direction of space defined by the locations of each pair of two smallest quantities of matter that the two objects comprise. The concept of a centripetal force acting between the gravity centers of objects m_1 and m_2 was invented by the clever physicist—recall the clever physicist from section 5—who is able to measure or calculate the position of a gravity center. Nature doesn't have that ability according to FLH. Her (enormous) power is that she can execute her law in the trillions of directions simultaneously, in any instant of time. It follows that *the U^{eq} region encompasses all the lines (directions) of interactions between the two objects*. (Approximately, for some energy U attached to the objects is neglected.)

At time t_1, the falling apple is two meters above the ground. At time t_2, it is one meter above the ground. The amount of energy U^{eq} contained in the apple-earth U^{eq} region decreased between t_1 and t_2 as the apple and the earth spontaneously advanced toward a greater degree of equilibrium, that is, toward a no-motion state in agreement with law Ia. Therefore, some equilibrium energy U^{eq} had to be released from U^{eq} region as illustrated in Fig. 9.

Conversely, lifting an apple causes an increase in the amount of U^{eq} contained in the apple-earth U^{eq} region. That is because lifting an apple leads to a decrease in the degree of equilibrium of the apple-earth system, as opposed to the falling apple scenario. An input of outside energy is required. The work done by a person lifting an apple is used to lift the apple and *not* to increase the gravitational-potential energy of the system. Potential energy does increase as energy U is absorbed from the surroundings of the U^{eq} region. The increase in the amount of equilibrium energy U^{eq} inside the apple-earth U^{eq} region is equivalent to the increase in Newtonian gravitational-potential energy. That will be discussed in section 27. In summary, the work done by the person lifting an apple up is used to lift the apple and not to increase the potential energy of the system.

Thus, the existence of energy U as a material substance, which in gravitational interactions reveals itself as the released/absorbed energy U, is concluded from the consideration of the results of the LIGO experiment and law Ia. LIGO results indicate that released energy spreads out across the universe in the form of gravitational waves (similar to the immaterial ripples in spacetime that spread out in GR). A fundamental property of energy U revealed in the LIGO is that it gravitationally interacts with mass objects. An emergent property of gravitational waves confirmed by LIGO observations is that they travel over cosmic distances at the speed of light c.

Material energy U is suggested to be the same physical entity as the energy released/absorbed in chemical reactions and phase transitions. The latter is normally called "binding energy," which corresponds to the amount of work required to take apart a stable system of atoms/molecules. Herein, that binding energy is called equilibrium energy (U^{eq}). A falling apple-earth system is moving toward a state of the lower energy as some material energy U^{eq} is released. The system is spontaneously moving toward a higher-degree-of-equilibrium configuration, that is, toward a no-motion state, which is nature's preferred state of matter. The formation of a molecule out of atoms and other molecules (a chemical synthesis reaction) also is spontaneous and results in the release of energy, normally referred to as the release of heat. The energy of a strongly bonded (i.e., in a high-equilibrium-state) molecule is less than the total energy of separated molecule ingredients. That the energy (dimension $[M \cdot L^2 \cdot T^{-2}]$) of a gravitationally interacting system decreases with increasing proximity of the objects that form the system is well known (Misner, Thorne and Wheeler[50]). According to the theory proposed here, that decrease results from the release of material energy U^{eq} (dimension [M]), as objects are advancing toward a higher degree of equilibrium.

Concerning phase transitions, the same applies, for instance, to ice formation (water freezing), the process in which water gets toward a higher degree of equilibrium while releasing latent heat, which means while releasing energy U^{eq}. The relation between heat and energy U will be discussed in more detail in section 28.

A chemical dissociation reaction involves separating atoms or molecules. That corresponds to the weakening of the bonds that bind atoms and molecules, which is associated with an input of heat energy to the molecule. Lifting an apple also requires an input of material energy to the apple-earth system according to the same physical principle: In both cases, the degree of nonequilibrium in the system increases, which is against progressing toward the natural state of matter. According to law Ia, a system of interacting objects cannot be brought toward nonequilibrium without an input (absorption) of energy U. That applies to gravitational interactions as well as chemical reactions and phase transitions. For instance, hydrogen in water cannot separate from oxygen without an energy input, which means that the degree of nonequilibrium cannot *spontaneously* increase as nature does spontaneous work, via the SLT or one of the SLT's aspects (law II or law III), only in a process toward a-higher-degree-of-equilibrium.

The suggestion that heat energy released/absorbed in chemical reactions and phase transitions is the same as energy U released/absorbed in gravitational interactions has been made based on the analogy between progressing toward equilibrium/nonequilibrium in the gravitational and molecular systems. That proposition is supported by FLH for nature is unable to make decisions about whether to release or absorb a specific kind of energy depending on the character of interaction. In other words, it is suggested that only one energy (U) exists as a fundamental material substance. The coexistence of, and nature's deciding upon the choice of one out of two or more energy types to be released/absorbed would contravene FLH.

Assuming that gravitational interactions and chemical reactions are universal, absorbing energy U has to be allowed for throughout the universe. Consequently, the entire universe is expected to be filled with energy U including interstellar space, the interiors of mass-energy objects, molecules, atoms and nuclei. As suggested earlier in this section, the results of LIGO experiment support that expectation as the gravitational wave, a very small part of which was detected at the LIGO site, had to travel radially through a very substantial part of the universe. The PSR 1913+16 data suggest that energy is released constantly from systems of gravitationally interacting objects, which further supports the notion that material energy U fills the entire universe. That leads to the conclusion that a perfect vacuum (space without matter) doesn't exist. The entire universe down to the internucleon space is expected to be filled with trillions of interfering gravitational waves, that is, with the waves of energy U (matter) the majority of which are so miniscule that, in all likelihood, we'll never be able to detect them.

Because material energy U gravitationally interacts with mass objects, its density (D^U) is expected to increase closer to objects, which is schematically illustrated in Fig. 8 by shading, just like the density of air increases closer to the earth. In this, a supposition is made that material energy U gravitationally behaves the same as a material fluid that is pulled by mass objects. As energy U is material and gravitationally interacts with mass objects, its motion in the G-P field is expected to be subject to the real force of inertia.

Fig. 9 illustrates an important aspect of law Ia: If, for whatever reason, boundary conditions are such that energy U^{eq} cannot be released from a system of interacting objects, nature won't do any spontaneous work on that system, and no motion resulting from a toward-the-equilibrium interaction will occur. This means that no quantitative (observable) effect of gravitational interaction will be generated. Such conditions would exist where the average density of energy U inside U^{eq} region (denoted D^U_{inside}) equals the average density of U outside the region (denoted D^U_{outside}). Then, a release of energy U^{eq} would be prevented by the SLT not being executed for energy cannot spontaneously move into a region with equal (or higher) average energy density. If, for whatever reason, $D^U_{\text{outside}} > D^U_{\text{inside}}$, the degree of nonequilibrium in the system would increase, with the objects moving, or attempting to move, away from each other. Under the most commonly studied scenarios of interacting objects, it is that $D^U_{\text{inside}} > D^U_{\text{outside}}$ for D^U_{inside} is set by the high densities of energy attached to objects by gravitational pull. That can be appreciated from an examination of Fig. 9.

Another important aspect of law Ia is that in commonly studied gravitational interaction cases, the difference between densities D^U_{inside} and D^U_{outside} is expected to increase as the distance between two interacting objects decreases. This can be concluded from an examination of Fig. 9, which schematically illustrates the density of equilibrium energy U^{eq} increasing with a decrease in the distance between the objects. The average densities of energy U inside and outside U^{eq} region determine the gradient under which the SLT is executed while releasing energy U^{eq}. The gradient controls the rate of the output of energy U^{eq} from U^{eq} region. The larger is the difference in energy densities, the larger is the gradient, which implies that energy U^{eq} is being released faster.

Based on this consideration, an acceleration of an object in free fall toward another object is expected. If an object is in motion with no energy release, no acceleration of the object is expected.

The above discussion suggests the following relation applicable to spontaneous work by nature applied to a free fall of an object,

$$\begin{aligned}&\textit{the rate of spontaneous work} \equiv \textit{the rate of released energy } U^{\mathrm{eq}}\\&\Rightarrow \textit{acceleration}\\&\Rightarrow \textit{increase in the degree of equlibrium of a system}\end{aligned} \tag{19}$$

where ≡ means "equivalent to" and ⇒ means "a consequence of."

As energy U^{eq} is released in the free fall of two gravitationally interacting objects, a halo of released energy is expected to form around the system (Fig. 9), and a condition of nonequilibrium in the distribution of energy U outside the system is expected to arise. That is because the average density of U has to increase adjacent to the boundary of the U^{eq} region as U^{eq} is released. Owing to the execution of the SLT, the released energy U has to move away from the system of two objects in free fall, in the form of a gravitational wave, thus advancing toward equilibrium in the distribution of energy U outside the system. That is the wave of energy that was inferred by Taylor and Weisberg.[51] Following GR terminology, they called it "gravitational radiation."

For convenience, the concept of "average energy density" has been used above to explain the motion of energy U^{eq} into or out of an U^{eq} region. As nature doesn't realize space and owns no weigh scales, she cannot realize density. In other words, the SLT cannot be executed based on "average" amounts of energy for averages have to be measured and calculated, neither of which nature is able to do. On the other hand, nature executes her laws in instants of space (section 4). Consequently, she simultaneously executes the SLT between each of the two smallest points in the system of objects, in each of the trillions of directions and in each instant of time, which effectively removes the need to measure and calculate averages.

Based on the above discussion, the energy-conservation law (law VII) can be clarified. The law states that *material* energy U cannot be created or destroyed. That law is exact for nature has no hidden sources of, or storage bins for material energy (matter). It is a requirement of FLH as operating such sources and bins would have to involve decision making. It is a local law for it applies to every point in space and in time. It also is a global law for it applies to the entire universe.

The conception of the universe completely filled with material energy *U* could be seen far-reaching. That is for one might ask the simple question: How is it possible that thousands of physicists haven't suggested the same over more than two thousand years of doing physics? Asking that question, however, would be rather inconsequential. The fact is that physicists and philosophers in the past pursued very hard the idea of "something" filling the entire universe. They called it either air, aether, spacetime, vacuum energy, quantum fields, dark fluid, or dark energy + dark matter + visible matter, depending on the time in history and the issue at hand.

The long discussion of this section states the fourth hypothesis underlying the proposed EE theory: *The entire universe is filled with physically-real material-energy U (matter)* (dimension [M]). A perfect vacuum doesn't exist.

23 The miniscule principle

It needs to be realised that something that appears miniscule to the human mind, isn't miniscule from the perspective of nature. That is because nature doesn't realize space and time, whence she has no concepts of a spatial size or a time period. Hence, she cannot have a concept of a "miniscule." It thus follows that in explaining physics at the fundamental level, miniscule effects shouldn't be neglected without considering all larger-scale consequences of the neglect. While introducing the assumption, "... magnitudes of the third and higher orders can be neglected ..." might lead to a simple and accurate description of a physical effect on a local scale, it can also lead to a misconception of the fundamental physics concepts underlying that effect and/or a failure of its description over large intervals of space and time.

The cumulative effects of miniscule phenomena can often be observed and measured. For instance, while the amount of energy *U* released as a gravitational wave/energy *U* in the formation of a single molecule might seem to be too imperceptible to consider, the cumulative effect resulting from a large number of molecules being formed can easily be observed and measured as the amount of released heat. Similar reasoning applies to the minuscule, on a local scale, release of material energy *U* by the photon that runs through the G-P field while interacting with mass objects. Over a few-meter distance, that release is expected to be absurdly small and unmeasurable. Over large cosmic distances, however, the effect of that release can be easily detected and measured. This refers to the cosmological redshift effect that will be discussed in section 68.

Recall Commander David Scott standing on the surface of the moon and dropping a feather. The magnitude of the gravitational wave generated by the falling feather had to be so absurdly small that, it all likelihood, we'll never be able to detect a wave of that size. Yet, the wave had to be there for nature is absolute and exact, and the effect of the fall was clearly observable and measurable: The feather fell over a period of about 1.3 seconds from a height of about 1.4 m. The very long time it took the feather to fall was the result of a finite speed of the motion of material energy *U*, that is, a finite speed of the gravitational wave. (That will be discussed in section 44). If it weren't for the finite speed of the release of energy *U*, the feather would fall down to the moon surface instantly. Thus, there seems to be no question that an absurdly minuscule effect can be associated with a well-pronounced effect. That suggestion will be called "*the miniscule principle*." Keeping it in mind will be crucial to the appreciation of some aspects of physics discussed in the remainder of the book.

24 Photon, electron, energy (matter)

Sunbathing on a beach browns my skin, that is, my skin gets burnt bombarded by photons. It would be irrational to believe that the burning of my skin is caused by immaterial (abstract) quanta of the electromagnetic field that the photons are normally assumed to comprise. A photon, which is a quantum of electromagnetic field, presents a mathematical construct that can only exist in the mind of a physicist. According to FLH, nature has no ability to realize and act upon mathematical constructs. The real photon—that is, the one that burned my skin—had to comprise physically real matter. It was suggested in section 22 that matter and material energy *U*—the energy that has to be material for it quantitatively affected the LIGO arms—are the same thing. Thus, *the photon is suggested to comprise energy U for it has to be material while there is no known matter other than energy U*. The coexistence of two kinds of matter would be in disagreement with FLH as nature would have to somehow measure/determine the properties of two kinds of matter to tell one from another and then to decide which set of her laws to use, while she is unable to make measurements and decisions.

As indicated previously, the release of energy by the photon gravitationally interacting with other objects can be suggested from the consideration of cosmological redshifts. The photon running through a G-P field gravitationally interacts with mass objects (i.e., with "distant masses"). Observations indicate that the photon loses more energy (dimension $[M \cdot L^{2} \cdot T^{-2}]$) the farther it travels. The abstract energy of the photon, which represents its ability to do work, decreases as the amount of material energy *U* (dimension [M]) the photon comprises decreases due to its release. (That is for the speed of the photon is approximately constant, while the abstract energy of the photon is defined by its kinetic energy.) The *release of material energy U* by the photon running through the universe is in agreement with Zwicky's tired light hypothesis[52] concerning cosmological redshifts. The *absorption of material energy U* by the photon interacting with mass objects is suggested based on the results of the Pound-Rebka experiment[53], which revealed that the photon absorbs energy in interaction with a massive object—the earth in that case.

Based on the discussion presented in section 22, it is suggested that the following most basic, intuitive concepts in physics; which include *matter*, *mass*, *energy*, and *mass-energy*; all refer to the same fundamental, physically-real substance, called energy *U* (dimension [M]). That suggestion appears to be rational for the mass of the electron, which is material, and the energy the photon comprises can be converted into each other.[h]

---

[h] The electron-positron pair production in the collision of two gamma photons, referred to as the Breit-Wheeler prediction, hasn't been experimentally confirmed yet. In this discussion, that prediction is assumed to be correct. The two-photon production in electron-positron collisions has been confirmed.

To summarize: The photon is expected to comprise energy U that burned my skin, for there is no other material substance known. That suggestion is supported by the results of Pound-Rebka experiment and the examination of cosmological redshifts. Those experiments imply that the photons release or absorb their energies into/from their surroundings as they move through the G-P field, thus interacting with mass objects. The released/absorbed energy U of the photon has to be the matter the photon comprises. That resolves the suntanning paradox. It is therefore suggested that matter (mass)—for instance, the mass of the electron—and energy U are the same material and physically-real substance. The conclusion has to be that all matter, that is, all electrons and protons, comprises energy U (dimension [M]), including also the neutron that can decay spontaneously into the proton, the electron and the antineutrino. It follows that the antineutrino is expected to be a material particle as well.

The photon, as a quantum of material energy U, is expected to interact gravitationally with mass objects, such as stars, stones and planets. That interaction was first confirmed in the light-bending-by-the-sun experiment carried out in 1919 in support of GR. The photon is expected to maintain its composition by self-gravity for there are no other known forces that could be acting inside the photon.

The material energy U the photon comprises is labeled U_{photon}. The material energy U that the cores of the electron and the proton comprise is labeled U_{solid}. The core of a particle is schematically illustrated by the black spot in Fig. 8. The electron and the proton (including the neutron) will be referred to as U_{solid} particles. A key difference between a U_{solid} particle and a photon is that the former appears not to release energy U except in collisions.[i] In the case of the photon comprising U_{photon}, the situation is different. Energy

---

[i] The electron and the proton may be releasing energy U regardless of interaction with other mass objects at the rates that cannot be detected in the time frame available to a human observer. As the execution of the SLT by nature is expected to enforce the release of energy U into the surroundings from any elevated-energy object, the supposition that the proton and the electron release energy would be consistent with the SLT being an absolute law of nature. At this point in the book, the possibility that the proton and the electron release their energies at some extremely low (unobservable) rates is of little consequence. However, it will become essential when the life of a star is examined in section 65.

U is released or absorbed by the photon as the degree of equilibrium between the photon and other objects changes owing to the photon's gravitational interactions. That appears to be a well-substantiated notion owing to the results of cosmological redshift measurements and the Pound-Rebka experiment results.

The third state of material energy U, labeled U_{fluid}, is "fluid-like energy," which fills the entire universe between U_{solid} particles. It includes energy U_{photon}. Thus, it is suggested that the universe is entirely filled with matter, which is energy U discussed in section 22. In this suggestion, an implicit assumption is made that energy U_{fluid} is non-corpuscular. Energy U_{fluid} also includes the energy that is pulled by gravity and attached to mass objects, as shown in Fig. 8. Since energy U_{fluid} comprises no particles, it isn't expected to have a property called "intermolecular friction." Consequently, its resistance to motion is expected to be due entirely to the inertial effect. That is rather similar to the resistance to motion of a superfluid substance.

Thus, the energy U_{photon} that the photon comprises is the same as energy U_{fluid} except that, inside the photon, relatively high density of U is maintained by self-gravity. As will be discussed in section 33, the photon is expected to be formed in a high-density U_{fluid} environment.

The above discussion allows for the clarification of the concept of a gravitational mass-energy object (g-object), which is critical to both postulate (1) and gravitational interactions in general: A g-object comprises U_{solid} particle cores (the cores of the electrons, protons, and neutrons) and energy U_{fluid} that is attached by gravity to the U_{solid} particles inside and outside of the single-mass objects that a g-object comprises. The U_{solid} cores of the electron, the proton and the neutron are finite in size. Consequently, those particles as well as their arrangements (atoms, molecules, stones, stars, etc.) *must have well-defined ranges of gravity, which means well-defined G-P fields of finite size.*

The electron and the proton have been suggested to have "core" parts that comprise energy U_{solid}. Since the existence of material energy U_{fluid} isn't recognized in current physics, it is a fair guess that the typical density of U_{fluid} is many orders of magnitude lower than the density of U_{solid} cores. That is because energy U_{fluid} has not been experimentally detected, so far. That, however, is not entirely true because, as discussed before, it is suggested that energy U_{fluid} was indirectly detected in the LIGO experiment as well as in chemical reactions (as heat). Also, as will be suggested in section 41, waves of energy U have been indirectly observed in double-slit experiments.

25 Work by nature

A physicist is looking at an apple hanging motionless from a tree. S/he realizes that because the apple isn't in motion, the degree of equilibrium of the apple-earth system cannot be changing, which is implied by relation (19). According to law Ia stated in section 3, that means that energy U^{eq} contained in the apple-earth U^{eq} region is not being released. Later, the apple falls and collides with the earth. The physicist wonders why the apple didn't fall to the ground instantly. Looking at relation (19), s/he conjectures that it didn't because equilibrium energy U^{eq} cannot be released instantly from a U^{eq} region. That is for the speed of the motion of material energy U, denoted c^{U}, is finite as will be suggested in section 44. Realizing that the speed of the motion of material energy U is finite, *the physicist conjectures that the speed of an object in free fall is restricted by the speed of the motion of material energy U being released from* U^{eq} *region.* The physicist is fairly certain of her/his conjecture for there seems to be nothing else out there that could affect the speed of a free-falling object (air resistance and similar effects are neglected). It follows that

$$\begin{gathered} speed\ of\ an\ object\ in\ free\ fall \propto \\ the\ rate\ of\ the\ release\ of\ energy\ U^{\text{eq}} = f(c^{\text{U}}). \end{gathered} \tag{20}$$

As pointed out in section 22, the rate of energy U^{eq} release is expected to increase as the distance between two objects in free fall toward each other decreases, which leads to the acceleration of an object in free fall toward another object. That is in agreement with the results of the Galileo experiments.

Relation (20) reiterates why the feather dropped by Commander Scott didn't fall to the moon surface instantly. The long time of the fall (about 1.3 seconds) was the result of the long time needed for energy U^{eq} to be released from the feather-moon U^{eq} region.[j]

Nature cannot do any *net work* since no work can be done out of nothing. That means that the work by nature has to be conserved with respect to both the works done toward a greater degree of equilibrium and toward a greater degree of nonequilibrium. That states the nature's-work-conservation law,

$$total\ (spontaneous + contra)\ work\ by\ nature = 0. \qquad (21)$$

This relation requires an explanation. As pointed out in section 22, material energy U^{eq} released during a free fall is expected to spread throughout the universe in the form of a gravitational wave. As the universe should be considered infinite in space (section 4), a gravitational wave, which comprises material energy U released from a U^{eq} region, will eventually be absorbed in (forced into) trillions of other U^{eq} regions. The absorption of energy U will be associated with the "contra" work done by nature in each of the trillions U^{eq} regions. A contra work done by nature, equivalent to an increase in the amount of U^{eq} *absorbed* inside the trillions U^{eq} regions, has to be equal to the amount of spontaneous work done, which is equivalent to the amount of energy U *released* from a U^{eq} region. That is enforced by law Ia.

After gravitational waves were released in response to the spontaneous work done by nature on the two free-falling neutron stars, it took a very long time before a small portion of those waves was absorbed and contra work done inside the LIGO arms-earth U^{eq} region. Because nature doesn't realize time, relation (21) was satisfied from her perspective: *As far as she was concerned, the spontaneous and the contra works were carried out in the same instant of time for she doesn't know of any other instants of time (time doesn't exist in nature's eyes). It follows that the material-energy-conservation law was enforced regardless of time and space*. But, relation (21) was also satisfied from the perspective of

---

[j] As will be discussed in section 43, the speed of the motion of energy U_{fluid}, which is the speed of gravitational wave, is the same as the speed of light c. That is the speed of the photon comprising energy U_{photon}. Neither energy U_{fluid} nor energy U_{photon}, however, is expected to accelerate to speed c^{U} (or c) instantly. That could be acceptable in a mathematical model only.

the physicist for during the long wave-travel time (about 1.3 billion years) spontaneous work by nature was being done as the wave of material energy U was moving in the process of reducing the degree of nonequilibrium in the distribution of energy U across a portion of the universe. That work by nature executing the SLT was spontaneous and in the same amount as the work done on the two neutron stars. The amount of energy U^{eq} released from the stars' site was equal to the amount of energy U that travelled as a gravitational wave. Owing to the nature's-work-conservation law, that suggests that the released energy U is equivalent to the work by nature since the work done on the two stars had to be equal to the work done by the wave. (Nature cannot possibly know it but the physicist can for s/he owns space and time). Upon arriving at the LIGO site, a small part of the spontaneous work by nature turned into contra work as energy was forced into the earth-LIGO arms U^{eq} region. Both the work by nature and energy U were conserved. It appears therefore rational to suggest that the work by nature and material energy U are equivalent.

The amounts of spontaneous and contra works by nature have to be equivalent to the amounts of energy U released and absorbed, respectively, because the work done by nature and the amount of energy U have to be conserved. (Keep in mind that the amount of energy released from a U^{eq} region at a source of a gravitational wave and absorbed by another U^{eq} region are expressed in the dimension of mass [M], while the amount of work done by nature is expressed in the dimension of work [$M \cdot L^2 \cdot T^{-2}$].) Therefore,

$$\begin{gathered} amount\ of\ spontaneous\ or\ contra\ work\ by\ nature \equiv \\ amount\ of\ released\ or\ absorbed\ material\ energy\ U\ (\propto \\ degree\ of\ equilibrium\ increase\ or\ decrease). \end{gathered} \tag{22}$$

In summary, there are two kinds of work that nature does. First, nature does *spontaneous work* in a toward-a-greater-degree-of-equilibrium process enforced by the execution of the SLT, law II or law III. That is the work associated with the release of energy U^{eq}. If boundary conditions are such that energy U^{eq} cannot be released, there can be no motion as nature is prevented from doing any work (law Ia). The work done in chemical synthesis and the work done on objects in free fall are examples of spontaneous work by nature. Second, nature does *contra work* in a toward-a-greater-degree-of-nonequilibrium process that is associated with the absorption of energy U. Examples of contra-work are chemical disassociation reactions (heat is absorbed) and the motions of LIGO arms forced by the motion of gravitational waves energy U^{eq} is absorbed).

In a spontaneous work done by nature, nature's engine is run by the SLT, the same as any man-made engine. However, in nature's work, energy U (dimension [M]) isn't turned into any other kind of energy for U has to be conserved. Thus, nature doesn't use up any energy U to carry out her work, and she doesn't rely on any external sources of work because such sources cannot exist. She just executes the SLT at the fundamental level. How does she do it isn't for us to know for one cannot know what lies beneath a principle that is a fundamental (irreducible) principle.

As previously stated, the standard view is that KE of an object in free fall ($KE = ½mv^2$) determines the ability of an object to do a certain amount of work. That view, however, applies only to the situation where an object which, being in internal equilibrium, moves as a whole with speed v in the frame of an observer. The amount of work that "an object can do" under that circumstance is equal to the amount the work that nature actually is doing on the object in the frame of the observer. That amount of work is much smaller than the amount that nature could potentially do using material energy U (matter) the object comprises. That amount is independent of the speed of an observer. For instance, let a piece of uranium ore with mass m be falling toward the earth with speed v. The amount of work that nature could *hypothetically* do via a phase transition process, matter-antimatter annihilation, or fission using the piece of uranium ore is many orders of magnitude greater than $½mv^2$, and independent of the speed of an observer. That, of course, was discovered by Einstein.

26 The motion-conservation law

The motion(momentum)-conservation law is well established and its validity has been well confirmed. The highlights of that law are: Motion (momentum) is a vector. The amount of motion is the amount of an object's linear or angular momentum. A system of objects is considered, which is an assembly of more than one single-mass objects. The motion-conservation law applies to a closed system. A closed system is one in which no change in the amount of mass occurs and the *net* external forces acting on the system are zero. Objects emit and absorb photons (matter), which means that a perfectly closed system with respect to motion may not exist. Note that all objects in motion in any closed system located in a G-P field are subject to real forces of inertia, however, those forces exactly balance out in a closed system owing to the action-reaction law.

The motion-conservation law cannot hold the status of a law of nature for motion is defined in terms of the amounts of time, space, mass and direction that nature is unable to measure and realize. Thus, motion isn't a fundamental physical entity such as material energy U and the motion-conservation law needs to be looked at from the perspective of the physicist.

Work is a key physics concept that belongs to the physicist. Any work means motion. Since material energy U is equivalent to the work done by nature (section 25) and must be conserved, the work done by nature has to be conserved as well (Eq. 23). It follows that motion has to be conserved. Thus, the motion-conservation law appears to be a derivative of the material-energy-conservation law referred to in section 22. As a result, the motion-conservations law is expected to be absolute and exact, the same as the material-energy-conservation law. The purpose of the above discussion isn't to re-invent the motion-conservation law. The goal is to imply that the motion-conservation is absolute and exact.

A g-object represents well a closed system for the purpose of the motion-conservation law. That is for the net external forces acting on the components of a g-object are zero. However, it isn't a perfectly closed system for a change in the amount of matter associated with the motion of the photons across its boundary is expected to be very small but not zero.

27 Gravitational-potential energy and function *K*

Misner, Thorne and Wheeler concluded, in their famous book, that the gravitational-potential energy (*PE*), cannot be localized. That conclusion may be seen strange as *PE* represents one of the most fundamental concepts in modern physics. Yet, that strangeness doesn't come as a surprise as typically no attention is paid to what *PE* actually is. *PE* is the ability to do an amount of work. It is a mathematical quantity. In physical terms, it doesn't exist. It is only the physicist who, in her/his mind, can calculate how much work can be done using a stretched spring or a falling apple. *Nature has neither any need nor interest in making such calculations*. The proper comment on the Misner-Thorne-Wheeler's conclusion is: Gravitational-potential energy *PE* (dimension $[M \cdot L^2 \cdot T^{-2}]$) cannot be observed/localized for it exists in the physicist's mind only. Abstract ideas cannot be observed/localized.

The concept of energy U (section 22) clarifies the *PE* location problem. The material equilibrium energy U^{eq} (dimension [M]) is numerically equivalent to the *PE* (dimension $[M \cdot L^2 \cdot T^{-2}]$). That is for the amount of U^{eq} released in gravitational interaction between two objects is equivalent to the amount of work done by nature, in accordance with relation (22). That means that it is equivalent to the loss of ability to do potential work. It follows that the abstract *PE* is detectable and localizable *via* that equivalence.

There is a simple reasoning underlying the above suggestion. Consider a system of two objects subject to direct gravitational interaction. The higher is the degree of equilibrium of the system, the less work the system can potentially do for there is less *PE* in the system, and vice versa. As the constituents of the system are brought apart, the amount of U^{eq} contained in the system—more accurately, contained in the U^{eq} region of the system—increases as material energy U is absorbed from the surroundings to become energy U^{eq}. Then, the degree of nonequilibrium of the system increases (law Ia), and the amount of work that the system can potentially do increases as well. Thus, a release or absorption of equilibrium energy U^{eq} is associated with a decrease or increase, respectively, in the amount of work that the system can do. That amount is the calculated amount of *PE* (an abstract, mathematical entity), which reflects the degree of the equilibrium of a system. Abstract *PE* can therefore be, in a manner of speaking, detected and localized via the U^{eq} - *PE* equivalence, keeping in mind that *PE* cannot be localized in a real sense, for it exists only in the physicist's mind. Material energy U^{eq} is stored (localized) in the U^{eq} region of two interacting objects.

Concerning a single-mass object, its gravitational energy PE (dimension [M·L^{2}·T^{-2}]) and material energy U^{eq} (dimension [M]) are also equivalent, with energy U^{eq} stored inside the object. It is stored in the U^{eq} regions located inside the core an object. Under that circumstance, PE doesn't depend on the distance to (i.e., the degree of equilibrium with) another object. There is no "another" object in relation to a single-mass object, which doesn't directly interact with any other object, except for distant masses.

Function K was introduced in proportionality (8) to account for a finite range of gravity. The arguments of function K, d_L and d, were selected such that the strength of force F_g^δ would become zero at the maximum direct interaction distance d_L (at $d = d_L$). That selection presents a shortcoming of function K, as it implies that nature realizes distance. The argument for the existence of energy U (section 22) allows for selecting an alternative argument of function K. Energy U^{eq} is the equilibrium energy that is released or absorbed in gravitational interactions. It can be used as an argument of function K, replacing Eqs. (9) and (10) with

$$F_g^\delta = \frac{1}{t_L^2 D}\left(\frac{1}{2}\frac{m_1}{d}\frac{m_2}{d} + \frac{1}{2}\frac{m_2}{d}\frac{m_1}{d}\right) K(U^{\text{eq}}). \tag{23}$$

A state of equilibrium between two objects doesn't change if the objects don't interact directly, and the direct force of gravity (force F_g^δ) is zero. That happens at distances between the objects greater than d_L, such that the G-P fields of the objects are separated. When two objects collide and the system comes to be in equilibrium (in a no-motion) state, the value of K is expected to have its highest value, which can be assumed as $K = 1$. Thus, similar to the simplest form of function $K(d_L, d)$ given in Eq. (10), the simplest form of $K(U^{\text{eq}})$ that satisfies the above conditions is

$$K(U^{\text{eq}}) = k\frac{U_{max}^{\text{eq}} - U_{\text{released}}^{\text{eq}}}{U_{max}^{\text{eq}}}, \tag{24}$$

where U_{max}^{eq} is the maximum equilibrium energy contained in a system of two objects, at zero equilibrium ($U_{max}^{\text{eq}} - U_{\text{released}}^{\text{eq}} = 0$), and $U_{\text{released}}^{\text{eq}}$ is the amount of equilibrium energy U^{eq} released in free fall since the state of zero-equilibrium.

CHAPTER V – HEAT ENERGY

28 Heat and temperature

Heat is normally interpreted as the kinetic energy (KE) of jiggling-vibrating atoms and molecules in transfer from a hotter to a cooler system with the measurement of temperature (T) interpreted as the measurement of the average KE of atoms/molecules in substance. As discussed in section 21, however, KE is a calculated (abstract) quantity. It appears inconceivable that an amount of work done, which is facilitated by a given amount of "heat," could be set by abstract KE that exists in physicist's mind only. The existence of physically-real material energy U suggested in section 22 allows for more meaningful interpretations of the concepts of heat and temperature measurement: *Heat is material energy* U_{fluid}, and *the measurement of temperature is the measurement of the density of energy* U_{fluid} ($D^{U_{\text{fluid}}}$). Those are incomplete interpretations for material energy U_{solid} is omitted from the consideration. One of the advantages of the proposed interpretation of temperature is the replacement of the temperature dimension [Θ] with the dimension of energy density [ML^{-3}].

As mercury level rises in a thermometer, it is a result of nature executing her laws. That level cannot be set by the average KE of atoms and molecules for nature is unable to calculate or otherwise determine the amount of average KE. Yet, it is evident that an input of heat into an object increases the frequencies of the vibrations of the particles the object comprises. In support of the suggestion that heat is energy U_{fluid} consider this: Forcing energy U_{fluid} (heat) into an object, e.g., heating a stone, increases density $D^{U_{\text{fluid}}}$ inside the object. Consequently, from the assumption that heating an object increases the frequencies of the vibrations of the particles the object comprises (i.e., increases their average KE as well as object's temperature), the following relation is suggested,

$$T \propto D^{U_{\text{fluid}}} (\propto \overline{KE}), \qquad (25)$$

where $\overline{KE}$ is the average KE of vibrating particles in a substance, the temperature of which is measured. As density $D^{U_{\text{fluid}}}$ outside a thermometer increases, heat (herein interpreted as energy U_{fluid}) is forced into the inside of the thermometer by the execution of the SLT in toward-a-greater-degree-of-equilibrium process. Mercury in the thermometer is expanding because a forced input of energy U—to become energy U^{eq} as contra work is done by nature—advances the thermometer toward a lesser degree of internal equilibrium, that is, toward weaker mercury binding forces and larger molecule separation distances.

Since heat is interpreted to be the same as material energy U_{fluid}, heat is material. Heat being a material substance was suggested by Carnot[54]. Thomson[55] and Maxwell[56] refuted Carnot's suggestion arguing that there was no analogy between heat flow and waterfall flow engines, and that material substance cannot be converted into immaterial mechanical energy. The Thomson-Maxwell arguments apparently settled the issue in the 19th century, and the concept of heat being an immaterial construct responsible for the transfer of abstract energy has been widely embraced ever since.

The proposal that material energy U_{fluid} and heat are the same thing implies that Carnot was correct in suggesting that heat was material. Maxwell's strong argument that material heat cannot be converted into immaterial work was, of course, correct. According to FLH, however, that argument isn't relevant: Energy (dimension [M]) conversion into work isn't possible in any case for energy U has to be conserved, which also means that it cannot be converted into immaterial work (dimension $[M{\cdot}L^2{\cdot}T^{-2}]$), and vice versa.

The preceding discussion leads to a simple explanation of the concept of absolute zero. Temperature is density $D^{U_{fluid}}$. As energy U_{fluid} fills the entire universe (section 22), its density cannot be zero. It follows that a minimum temperature, corresponding to density $D^{U_{fluid}} = 0$ and identified with the condition of *absolute-zero temperature*, exists in hypothetical sense, while it can never be reached for $D^{U_{fluid}} = 0$ can never be reached. The underlying premise is that space empty of matter, cannot exist, as suggested previously in section 22. That appears to be consistent with FLH as having empty space would mean that there are parts of the universe in which nature doesn't exist: There wouldn't be any matter in those parts, whence there wouldn't be any laws of nature. Nature comprises matter and her laws. She cannot exist without herself such that no part of the universe can be empty of matter.

An object is heated. As temperature, which is equivalent to $D^{U_{\text{fluid}}}$, inside an object increases, the amount of material energy U contained inside the object increases as well. Consequently, the internal PE of the object is expected to increase. Heating an object will lead to an increase in the amount of equilibrium energy U^{eq} contained in the molecular U^{eq} regions inside the object which, in turn, will lead to the weakening of the gravitational forces inside the object.

29 Steam and waterfall engines

In terms of fundamental physics, an engine is a man-made device that harnesses work by nature and releases-transfers that work such that it can be turned into useful work (into "motive power"). In man-made engines, the spontaneous work by nature means the execution of the SLT. It always involves the release of material energy U^{eq} in accordance with law Ia.

Thomson's argument against heat being a material substance refers to Carnot's suggestion that a steam engine is analogous to a waterflow engine, which implies that material heat flow caused by a difference in temperatures is somehow analogous to waterflow caused by a differential water head. The two kinds of engines are illustrated in Figs. 10 and 11.

In a steam engine, the difference between the temperature of hot gas in the pipe running through water in the boiler (Fig. 10) and the temperature of boiler's water results in spontaneous work done by nature in a toward-the-equilibrium process. She executes the SLT across the pipe wall. Nature's work forces the motion of heat (U_{fluid}) *that contains no "heavy"* U_{solid} *particles* (i.e., no electrons, protons, or neutrons) such that the inertia forces opposing the motion of heat are expected to be extremely small. Hence, the amount of work done by nature in moving heat across the pipe wall must be extremely small as well, in accordance with the action-reaction law. That work by nature isn't transformed into motive power as Carnot may have thought. It just brings the water in the boiler toward a greater degree of nonequilibrium as the binding (intermolecular) forces in the water get weaker due to the forced absorption of material energy U_{fluid} (heat).

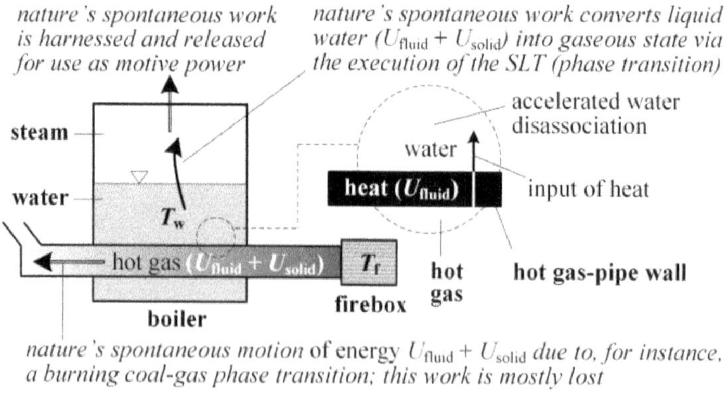

nature's spontaneous work is harnessed and released for use as motive power
nature's spontaneous work converts liquid water ($U_{fluid} + U_{solid}$) into gaseous state via the execution of the SLT (phase transition)
accelerated water disassociation
steam
water
water
heat (U_{fluid})
input of heat
T_w
hot gas ($U_{fluid} + U_{solid}$)
T_f
hot gas
hot gas-pipe wall
firebox
boiler
nature's spontaneous motion of energy $U_{fluid} + U_{solid}$ due to, for instance, a burning coal-gas phase transition; this work is mostly lost

Fig. 10 Schematics of steam engine working according to FLH. Material heat (energy U_{fluid}) is forced across the hot-gas-pipe wall (a result of the execution of the SLT). The inflow of heat into boiler's water maintains a high rate of water-steam transition as temperature T_w, proportional to the density of U_{fluid} in water, is maintained high. The transition converts boiler's water into steam to become useful work. No U_{solid} particles (electrons, protons, neutrons) are transferred across the gas-pipe wall. U_{solid} particles are transferred from boiler's water to steam. Nature's work on the motion of U_{solid} particles in the gas pipe is lost. Nature's work on the motion of U_{fluid} in the pipe is partly lost as some heat (U_{fluid}) is transferred to boiler's water.

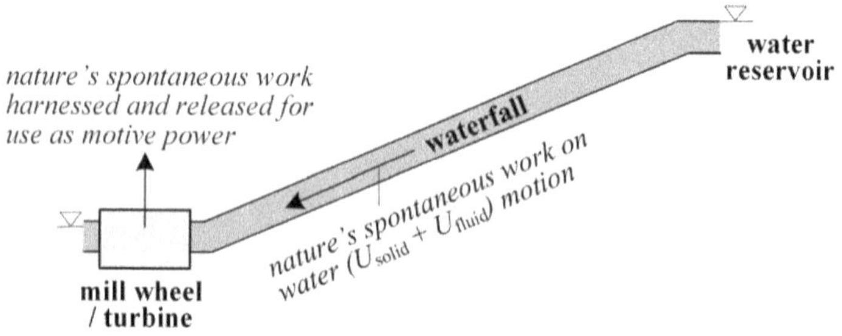

water reservoir
nature's spontaneous work harnessed and released for use as motive power
waterfall
nature's spontaneous work on water ($U_{solid} + U_{fluid}$) motion
mill wheel / turbine

Fig. 11 Schematics of waterfall engine. Spontaneous work is done by nature on the motion of water, comprising U_{solid} + U_{fluid}, along an unconfined waterfall stream or penstock. That work (minus friction losses) becomes useful work. No transfer of heat or a phase transition are involved.

The weakening of intermolecular forces due to contra work by nature leads to a high rate of phase transition as the boiler's water turn into gaseous water in the process that generates motive power. In the transition, *the work by nature results in the release of water molecules, which comprise* $U_{solid} + U_{fluid}$, *such that the amount of work by nature is no longer "extremely small" for it involves the motion of "heavy"* U_{solid} *particle cores*. This is the work that is transformed into useful work, most often, by the motion of a piston. There is no momentum of substance passing through the gas-pipe wall while there is a lot of momentum passing from the boiler's water into the gaseous space water due to the masses of protons, neutrons and electrons contained in the cores of the particles.

The rather trivial point made above is that the work done by a person shoveling coal into the firebox of a locomotive (to keep gas temperature high) is ridiculously small as compared with the work needed to move the locomotive wheels. Furthermore, the work by nature in moving heat across the pipe wall is ridiculously small. The question has to be: "Who/what does the work to move the wheels"? The answer appears to be straightforward: It's the phase-transition work by nature inside the boiler, triggered by the rise in the temperature of boiler's water (by a conflict) for there is no one/nothing else out there that could move the locomotive. Let the temperature in the firebox become a thousand times normal. That wouldn't move the locomotive a millimeter if the broiler was dry. That means: No phase transition, no motive power. The ingenious contribution of the human mind is setting up the conditions under which nature's work is done. Those involve the motion of heat having negligible momentum across the hot-gas-pipe wall. It brings the toward a greater degree of nonequilibrium as a result of contra work done by nature. In other words, energy U from the hot gas pipe is forced into the boiler's water to become energy U^{eq}. As the nonequilibrium in the boiler's water increases, the water molecules are released from the boiler's water. The molecules are "heavy" for they contain the U_{solid} cores of particles. Their motions (momenta) are used to produce motive power.

In a waterfall engine (Fig. 11), the spontaneous work by nature to be turned into motive power doesn't involve the motion of heat, or phase transition. The amount of useful work equals the amount of work done by nature in eradicating nonequilibrium in the earth-water-reservoir system. The heavy U_{solid} particles flow down with their momenta converted into motive power.

Thomson's argument that the differences in temperatures and in water heads in the steam and waterfall engines, respectively, aren't analogous with respect to producing motive power was, of course, correct. Yet, that argument appears to be irrelevant in concluding whether heat is material or immaterial when the steam and waterfall engines are compared. That is because in neither case does heat transfer produce motive power.

CHAPTER VI – ELECTRIC FORCE

The ideas presented in this chapter are far-reaching and certainly controversial. They arise because of the suggested existence of material energy *U* (section 22). Keep in mind that the contents of this chapter have no bearings on the remainder of the book.

The ensuing discussion is intended to demonstrate how the simplicity of nature could be further argued for. The primary goal of this discussion is to show that the physical law for the electric force could perhaps be stated in terms of the three fundamental dimensions: mass, space, and time [M, L, T]. This is the same idea as the replacement of dimension [Θ] with [ML^{-3}] in section 28.

30 Gravitational versus electric field

It has been suggested that the G-P field introduced in section 8 is a physically-real information field, nonquantitative and perfectly homogeneous. To the contrary, the electric field, as normally understood, is quantitative for it comprises local values that vary with space and time. It is therefore nonhomogeneous. The primary difference between those two fields is that the G-P field represents an *invariant property* of mass, while the electric field represents an *emergent property* of electric interactions.

Let an interacting proton and electron be close to each other and located far away from any other charges. Then, let the proton be moved sufficiently far away such that it ceases to interact with the electron. The electric field surrounding the electron disappears. It appears impossible to detect the electric field around an electron without inserting a close-by charge.

Another problem with appreciating the meaning of the electric field is Coulomb's law,

$$F_e = k_e \frac{q_1 q_2}{d^2}. \tag{26}$$

Eq. (26) doesn't account for the mass of charged particles. It states that electric force would be generated between two charges even if the charges are massless. According to FLH, however, that cannot be true. *For two charges to interact quantitatively, which means to generate an observable effect, a mediating agent and a conflict with the execution of law III have to exist* (section 3). If the charges were massless, that is, comprised no matter, then no law other than law III could be executed such that there would be no conflict and thus no generation of an observable electric effect. However, all charged particles do have mass, and law II must be executed between any sufficiently-close-to-each-other charges. The simultaneous executions of laws II and III on two charges are expected to generate a conflict that underlies an observable electric interaction. Yes, it certainly is true that the strength of the direct force of gravity is miniscule in comparison with the strength of the electric force, when acting between two fundamental charged particles. Yet, that miniscule force appears to be the key to appreciating what it takes for law III to be executed in a quantitative way. Hence, it is suggested that the electric field isn't the agent that mediates electric interactions. That field appears to be an emergent property of electric interactions. The real agent mediating electric interactions is the G-P field that leads to a conflict and the resulting quantitative interactions. It follows that there appears to be only one physically-real force field in nature, the G-P field, just as there appears to be only one fundamental substance, which is called material energy U.

31 The core-attached energy interface

Consider the object illustrated in Fig. 8 and assume that its core, marked as the black circle, spins. The idea of the core of an object is extended to include both the U_{solid} cores of the electron and the proton (section 24), and the well-pronounced (visible) parts of the ordinary "large-scale" composite objects such as stars, the earth, and the LIGO arms. The cores of the ordinary objects have to consist, at least in part, of energy U_{solid} for we can discern those cores. On the other hand, we have no sense of energy U_{fluid} except for visible and infrared photons (section 5).

Concerning the spinning of the core of an object, an important assumption is made: *Material energy U_{fluid} attached by gravity to the spinning core of an object rotates in response to the spinning.* In more detail, it is assumed that the spin-generated local changes in the density of energy—in the area adjacent to the core-attached energy interface—result in some resistance to the sliding of U_{fluid}

along the interface.[k] The experiment that appears to support that assumption is the LIGO experiment, from which some resistance to sliding between the gravitational wave (energy U_{fluid} in motion) and the LIGO arms can be concluded. As the LIGO arms comprise the electrons, protons and neutrons, for the LIGO arms to move under a wave of energy U_{fluid} a resistance to the sliding of energy U_{fluid} over the arms is necessary. If the above assumption is correct, energy U_{fluid} attached to the spinning core of an object would be expected to rotate, whence it would be subject to the inertial resistance to motion due to the interaction of the rotating energy with distant masses.

32 Spin and charge

Thus far in this book, physical phenomena have been discussed in terms of Newtonian physics as four physical quantities have been considered: matter (i.e., energy U), space, time, and temperature. Those relate to the basic dimensions: kilogram, meter, second, and kelvin. It was suggested in section 28 that the measurement of the temperature is the measurement of the density of energy U_{fluid}. That eliminates kelvin from the list of basic dimensions.

Normally, five basic dimensions are considered: kilogram, meter, second, kelvin, and coulomb. Including coulomb appears to present a problem for it implies that charged particles comprise a special kind of matter. That isn't a rational supposition from the perspective of the universe that is filled with material energy U (section 22). If all matter comprises the same energy U, charged particles cannot constitute a separate substance. Those can only constitute a special state and/or form of interaction of energy U.

There appears to be no evidence for two kinds of matter to co-exist, one neutral and one electrically charged. The existence of more than one kind of matter would be untenable in view of FLH as this would imply nature to measure neutrality-charge and then decide which of nature's laws is to be executed based on the result

---

[k] As energy U_{fluid} comprises no molecules or atoms, it isn't expected to have a property called "internal friction" at the interface. However, the speed of motion of U_{fluid} is limited to the speed of light, as will be discussed later, in chapter VIII. As a consequence, local changes in the density of energy adjacent to the core-attached energy interface—which are generated by the spinning of a core—are expected to result in some resistance to the sliding of energy along the interface.

of the measurement. That would be a humanization of nature for she is unable to make measurements and decisions.

On the other hand, nature appears to be very simple if only one kind of matter exists (section 24). It includes corpuscular matter in the form of high-density energy U, that is, in the form of the cores of U_{solid} particles, the electron and the proton.[1] The cores of particles are "submerged" in energy U_{fluid} filling all inter-core-particle space in the form of energy U of a very low density. The corpuscular photons comprising U_{photon} also exist. In that portray of the universe, charged particles are expected to reflect a special form/behavior of material energy U.

In this section, a particle means either the electron or the proton. Neutrons are assumed to be formed by electron-proton pairs. Let there be an electron and a proton far away from any other objects. If a charged particle isn't a separate material substance, then there must exist some unknown form/behavior of material energy U that results in generating the emergent property of a particle, which we call charge.

The structure of a particle is illustrated in Fig. 12. It suggests that a significant charge cannot be a property of a particle without U_{solid} core (without the "rest mass" as will be discussed in the next paragraph). That is for in the electron-positron and the photon-photon collisions charges are destroyed and created, while U_{solid}

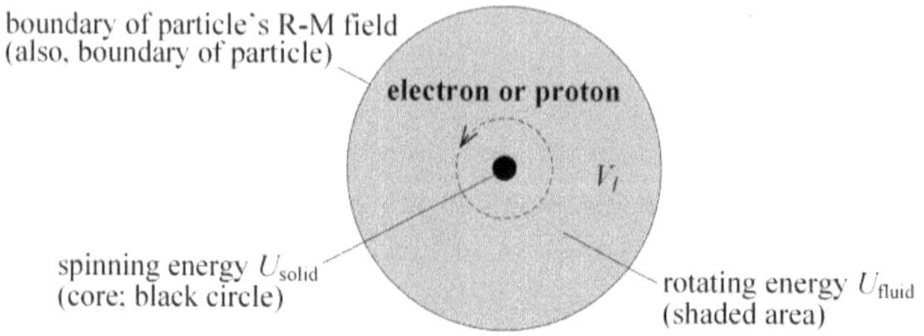

Fig. 12 Conceptual spinning of the U_{solid} cores of proton or electron. Energy U_{fluid} is pulled by gravity and attached to particle's core. As a core spins, the energy attached to it is subject to rotational motion. The rotational-motion (R-M) field of a particle, V_l, is approximately equal in extent to G-P field V.

[1] The standard view is that neutron cannot consist of a proton, an electron and antineutrino. It is a neutron decay that generates those three particles. Prior to the decay, neutron's internal structure comprises three point-like quarks and gluons.

cores of particles (their rest masses) are simultaneously destroyed and created. Note that as pointed out in footnote g, the prediction of the production of an electron-positron pair in the collision of two photons hasn't been experimentally confirmed. Herein, that prediction is assumed to be correct with the purpose of suggesting that a charge can be created.

As discussed in section 24, the observed cosmological redshifts and the Pound-Rebka experiment suggest that the photon releases or absorbs its energy due to gravitational interactions with other objects. Also, the photon is considered to have no rest mass, which is conjectured to mean that *it has no mass that cannot be released or absorbed in conjunction with its motion in G-P field.* In terms of material energy, the photon has no energy in the form of U_{solid}. To the contrary, the electron has rest mass in the form of U_{solid}. That energy is neither released nor absorbed as the electron is in motion through G-P field or, perhaps, it is released/absorbed at some extremely low rates (footnote h). In summary, if an electron is a component of a system in motion, its mass remains constant and independent of its motion or the motion of the system, at least to an extremely high degree of accuracy.

Thus, good chances are that the property of a particle, which is called charge, relates somehow to the core of a particle comprising energy U_{solid}. In the electron-positron collisions, both charges and U_{solid} cores disappear as energy U_{solid} is converted into energy U_{photon}. On the other hand, particles with no rest mass or particles with relatively very little rest mass (if any) such as photons and neutrinos—that is, particles comprising very little or no energy U_{solid}—don't carry measurable charges. That appears to support the supposition of a relation between the charge of a particle and its core comprising energy U_{solid}.

Let the U_{solid} cores of the particles shown in Fig. 12 be actually spinning, just like ordinary tops[m]. Energies U_{fluid} attached to the cores of the particles by gravity are expected to rotate due to the spinning of the cores (section 31). The cause of core spinning is

---

[m] In QT, particles cannot spin in classical sense for particles are point-like objects, and would have to spin with speeds greater than the speed of light to generate the actually measured magnetic effect. Assuming a particle to be point-like in a physical law, however, doesn't prove in any way that a particle cannot spin.

unknown. The orientations of the spinning axes are unknown. The speeds of core spinning are unknown. The sizes and shapes of particle cores are unknown. The only thing that can reasonably be assumed known is this: As the rotating energies U_{fluid} attached to the cores of two particles come into contact, a conflict is inflicted and a force between the particles has to be generated due to the spontaneous execution of a law of nature. That force, which will be called *the spin force* (F_S), has to act along the line that connects the two particles. That is for in quantitative execution of her laws, nature spontaneously drives two quantitatively interacting objects toward equilibrium (toward a no-motion state). This is explained in the following paragraphs.

It was assumed in section 31 that there is some resistance to the sliding of energy U_{fluid} along the core-attached-energy interface, whence the attached energy is subject to rotational motion. Then, a rotational-inertia-based-resistance force has to emerge, such that the action-reaction law is obeyed. The force of inertia is generated by the interaction of rotating energy U_{fluid} with distant masses.

The range of rotational motion of energy U_{fluid} is denoted "V_l". It defines the rotational-motion (R-M) field of the attached-to-the-core-and-rotating energy U_{fluid}. The R-M field is expected to be equal in extent to the G-P field of a particle. That is because the motion of rotating energy U_{fluid} cannot extend beyond the G-P field as energy U_{fluid} comprises no molecules or atoms, which suggests that it has no internal (intermolecular) friction. Contrary to the G-P field, the R-M field is an emergent field that is material and nonhomogeneous.

Let the two particles shown in Fig. 12 be located far away from any other particle/object. As long as the R-M fields of the particles are separated, the particles don't interact. Consistent with FLH, as the two R-M fields get connected, *a conflict is inflicted by the interference of two rotating energies U_{fluid}, and the particles start interacting due to energy rotations*. Nature, in the spontaneous execution of her law, drives the particles toward an equilibrium (i.e., toward a no-motion state). With respect to the equilibrium driven interactions of two particles, there are two possible no-motion (equilibrium) states. First, when the particles collide, and their quantitative interaction ceases. Second, when the R-M fields of the particles separate and the interaction between the particles ceases. In both cases, there would be no motion of the particles toward or away from each other due to the rotations of U_{fluid} energies. Equilibrium states would be reached in each case.

Thus, while quantitatively interacting, two particles have to be in motion either toward or away from each other such that they are advancing toward a no-motion state. Because nature operates in the directions of space (section 4) , the motion is expected to be along the straight-line connecting the interacting objects. As the particles interact and are in motion, there must be a force between them acting along that line, marked "line of force" in Figs. 13 and 14. That is the force that has been named the spin force, F_S. It acts on the particles in addition to the indirect force of gravity, F_g^{δ}.

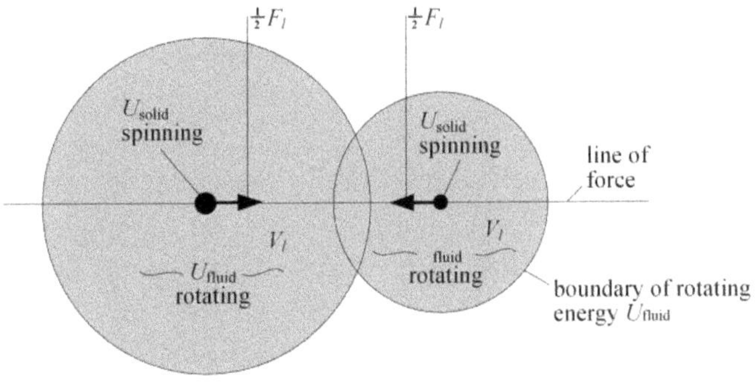

Fig. 13 Schematics: Spinning proton and electron generate an attractive force in linear motion toward equilibrium (toward a no-motionstate). The effect of the positions of rotation axes is neglected.

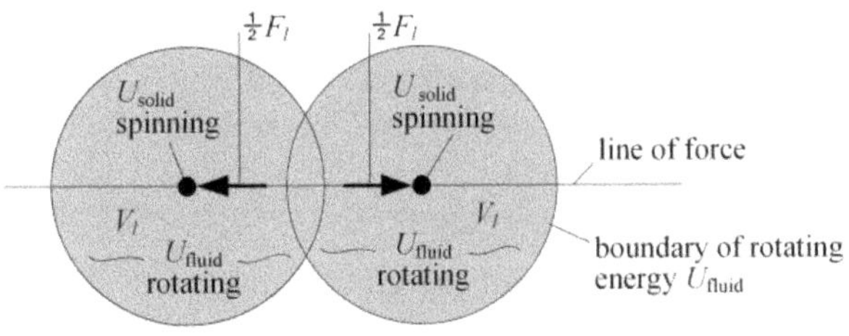

Fig. 14 Schematics: Two spinning and interacting electrons or protons generate a repulsive force in linear motion toward equilibrium (toward a no-motion). The effect of the positions of rotation axes is neglected.

The spin force is of the gravitational origin in the sense that it is associated with the gravitational interaction of rotating energies with distant masses. Furthermore, the rotating energies are bound to the particle cores by gravity. If the spin force is real and has been actually observed, it would have to be the electric force for there appears to be no other (known) force that could fit the bill. (It certainly wouldn't be the weak or the strong nuclear force.)

If two interacting particles with the same spin were moving toward each other, the system the particles form would experience an increasing amount of spin nonequilibrium. Consequently, it is expected that two same-spin particles are forced to move away from each other, that is, to move toward the state of a higher degree of equilibrium. A state of equilibrium would be reached as the R-M fields of the two particles separate, which would result in reaching a no-motion state. The opposite applies to two particles with opposite spins. In this case, nature's objective is the same: to reach equilibrium, which means to reach a no-motion state. That would be reached when the two particles collide and the motion due to the energy rotations ends.

The point that I am trying to make by presenting this section is that if the spin force exists indeed, it has to be the electric force for we don't know of other forces that could be considered here in lieu of the electric force. If so, the physical law for the electric force (26) can be expressed in terms physical quantities having the three fundamental dimensions: [M], [L] and [T].

Example 1: If the particle charges (q_1, q_2) are replaced with their angular momenta (l_1, l_2), the remaining components of the spin force equation would have the dimensions of length [L], time [T], dynamic viscosity [M/L·T], and kinematic viscosity [L^2/T].

Example 2: If the charges (q_1, q_2) are replaced with their moments of inertia (I_1, I_2) in Eq. (26), the remaining components of the spin force equation would have the dimensions of length [L], time [T], linear density [M/L], and an area [L^2].

The Example 1 is based on the postulate $l = \eta V_l$, where η is a dynamic viscosity. Example 2 is based on the postulate $I = CV_l$, where C is a linear mass density. In both cases, $V_l = V$ is the amount of gravitational potential. The coulomb wouldn't appear as a basic dimension in neither Example 1 nor in Example 2.

CHAPTER VII – OTHER MYSTERIES IN PHYSICS

33 Nucleus-electron interaction

According to FLH, physics at the classical level is no different from physics at the quantum level (section 2). In this and in section 36, I will contend that accepting FLH's implication that FLPs are the same at the classical and quantum levels offers the advantage of explaining the nucleus-electron and the sun-planet interactions in terms of the same fundamental principle.

The model of the atom to be discussed is the Bohr-Rutherford model with material energy U, equilibrium energy U^{eq} and law Ia added, and the SLT employed. A schematic nucleus-electron interaction is illustrated in Fig. 15. The system to be examined

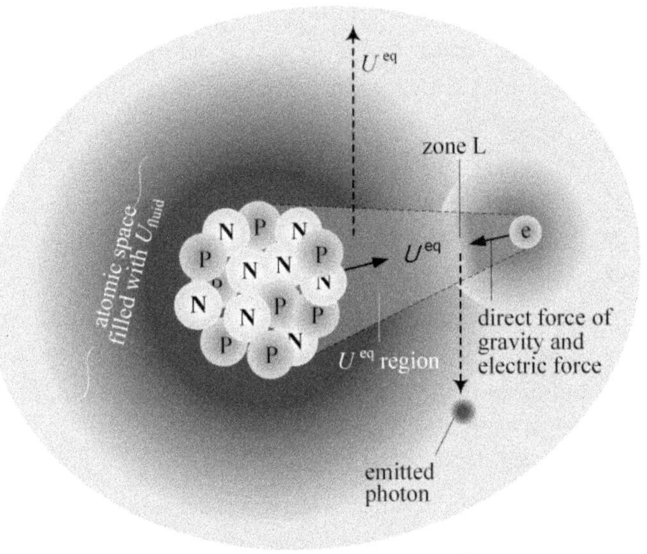

for electron to get closer to nucleus some energy U^{eq} has to be released from U^{eq} region (law Ia)

under stable-atom condition, electron cannot change its orbital (get closer to equilibrium) as no energy cannot be released from U^{eq} region

the nucleus remains stable as driving the nucleons apart requires more energy to be absorbed than released

Fig. 15 The direct force of gravity and the electric force drive the electron toward an equilibrium (a no-motion) state. Electron doesn't fall onto nucleus for its motion toward a state of graeter equilibrium is prevented as energy U cannot be released from U^{eq} region. That is because the average energy density in a U^{eq} region is about equal to the average energy density in the surrounding atomic space.

includes the electron, the nucleus, their U^{eq} region, and the atomic space surrounds the electron and the nucleus.

First, consider an atomic electron that is relatively far away from the nucleus. For the electron to get closer to the nucleus, the average density of energy U, $D^{U_{\mathrm{fluid}}}$, has to be higher in U^{eq} region than in the surrounding atomic space such that the SLT can be executed with a release of equilibrium energy U^{eq}. The execution of the SLT involves the motion (release) of energy U^{eq} away from the U^{eq} region as required by law Ia, because the electron and the nucleus are advancing toward equilibrium, that is, toward a no-motion state. Said another way, energy U^{eq} is released from U^{eq} region by the execution of the SLT as the electron, driven by the executions of laws II and III, is getting closer to the nucleus, that is, closer to the equilibrium of the nucleus-electron system. The electron falls toward the nucleus, just as an apple driven by law II falls toward the earth. There is, however, a crucial difference here. In the case of the electron, energy U^{eq} is released in the process of increasing *both* the degree of gravitational equilibrium and the degree of electric equilibrium. Let the (neutral) apple have a mass of the electron. The work done by nature in the case of a falling apple is $F_g^{\delta} d$. In the electron case, that work is $F_g^{\delta} d + F_e d$. The electric force is known to be orders of magnitude stronger than the direct force of gravity. From that, the amount of energy U released over distance d—which is equivalent to the rate of work done by nature (see section 25)—is concluded to be orders of magnitude larger in the case of the "falling" electron than in the case of the falling (neutral) apple of the same mass.

As the electron advances toward the nucleus, the amount of energy U^{eq} released becomes relatively large due to the electric interaction. That is expected to result in the average $D^{U_{\mathrm{fluid}}}$ inside U^{eq} region to decrease, at some point, to the average $D^{U_{\mathrm{fluid}}}$ in the adjacent atomic space. Under that circumstance, the electron can no longer get closer to the nucleus, as no energy can be released from U^{eq} region into the surrounding space, which has the same average $D^{U_{\mathrm{fluid}}}$. For the electron to get closer to the nucleus, that is, closer to the system's equilibrium, energy would have to be released from the system according to law Ia. Asking "how can one know that the average $D^{U_{\mathrm{fluid}}}$ in a U^{eq} region would actually decrease to the average $D^{U_{\text{fluid}}}$ in the surrounding space, and the release of energy would stop?" wouldn't be constructive. Simply, if a sufficient decrease in the average $D^{U_{\text{fluid}}}$ in U^{eq} region did not occur, the electron would keep advancing toward equilibrium, and would eventually fall onto the nucleus.

An electron that is not moving toward the nucleus isn't expected to be motionless for the distribution of density $D^{U_{\text{fluid}}}$ on both sides of the boundary of the U^{eq} region has to be nonuniform. As a consequence, energy U is expected to fluctuate across U^{eq} region boundary due to the execution of the SLT, as the electron and the nucleus are expected to be subject to some miniscule motions ("vibrations"). Because of the influence of neighboring atoms, the vibration pattern is likely to be highly complex. Furthermore, the electron doesn't need to be located along a specific radial direction extending from the nucleus for the forces between the electron and the nucleus would be independent of that direction. Said another way, the atomic electron is free to vibrate and move in any direction (subject perhaps to other constraints, if those exist) as long as it remains within approximately the same distance from the nucleus, that is to say, in the same "orbit."

Thus, if the release of U^{eq} from U^{eq} region is prevented, the electron cannot move toward the nucleus. However, the electron can still move closer to the nucleus if a photon, which comprises U_{photon}, is emitted from U^{eq} region. That is in agreement with relation (19) for the degree of equilibrium of the nucleus-electron system has to change if some energy U^{eq} (a photon) is released.

Fig. 15 also illustrates a possible explanation of why a stable atomic nucleus can exist. *All* nucleons are subject to gravitational interactions. A release of nucleons from the nucleus would mean moving toward a lesser degree of equilibrium. That would require absorption of energy U from the nucleus surroundings (just like in a chemical disassociation reaction). Roughly *half* of nucleons (all protons) are also subject to electric interactions. Releasing protons from a nucleus would mean advancing toward a higher degree of electric equilibrium, which would require a release of energy. On the balance, it is expected that for the nucleus to fall apart, the amount of energy that would have to be absorbed (all nucleons are flying apart) would have to be greater than the amount of energy that would be released (only protons are flying apart). Such a scenario, however, is not possible because energy U_{fluid} cannot spontaneously move from a region of a lower energy density (the atomic space around nucleus) into a region of a higher energy density (the interior of the nucleus). That is forbidden by the SLT. As a result, the nucleus remains stable.

34 Least-effort

As illustrated in Fig. 15, the photon emitted from atomic space is expected to be released from zone L for the gravity forces that bind material energy U^{eq} to the electron and the nucleus are weakest in that zone[n]. Thus, nature is removing ("emitting") the photon from the part of U^{eq} region that requires the least amount of removal effort. The least-effort proposal can be explained using a rubber balloon experiment. While being inflated, the balloon will eventually break at the spot where the rubber is the thinnest, all other factors being equal. If the thin-rubber spot is extremely small and barely thinner than the remainder of the balloon, the physicist will not be able to detect it. Therefore, the physicist will not be able to predict the location at which the balloon will break. Nature doesn't know the location of the break spot either. She will merely execute her laws in exactly the same manner (all other factors are equal) at each most miniscule spot of the balloon being inflated. The balloon will break at the thinnest rubber spot.

The above scenario concerns the notion of QT that the physicist cannot predict when, and from which atom, a photon will be emitted. Dehumanized nature doesn't know it either. The fact is that *she has no need to know it*. She executes her laws in exactly the same way at each electron-nucleus U^{eq} region. Rather than an uncertainty built into nature, that appears to be the inability of the physicist to identify *exact* initial and boundary conditions for each of the electron-nucleus U^{eq} regions, and to analyze and compare trillions of U^{eq} regions using physical laws. The physical laws are near-exact at best. Nature is unable to do those chores either, yet she causes the emission of the photon to occur from exactly the weakest zone L in a U^{eq} region in each instant of time. That is for

---

[n] To visualize zone L, consider two earth-like planets with their atmospheres overlapping. There is a zone between the planets where the air density is lowest while the forces that bind air to the planets are weakest. That is zone L.

she is absolute and exact, and can tirelessly and simultaneously perform trillions of tasks in each instant of time.

The least-effort rule is an emergent property of nature executing her laws. The background to that rule is nature executing her laws always in exactly the same way in all directions of space and in all instants of time. And, the least-effort conflict always "wins."

35 GR time dilation

Einstein predicted gravitational (or GR) time dilation from the principle of relativity, the assumed equivalence of a uniformly accelerated frame of reference, and a gravitational field.[57] The prediction indicated that a clock would tick faster the farther from a massive object it was located. (That often is stated as, "The lower is the gravitational potential, the flow of time is slower.") Herein, gravitational time dilation is concluded from the consideration of the release of energy from U^{eq} region, with reference to Fig. 16.

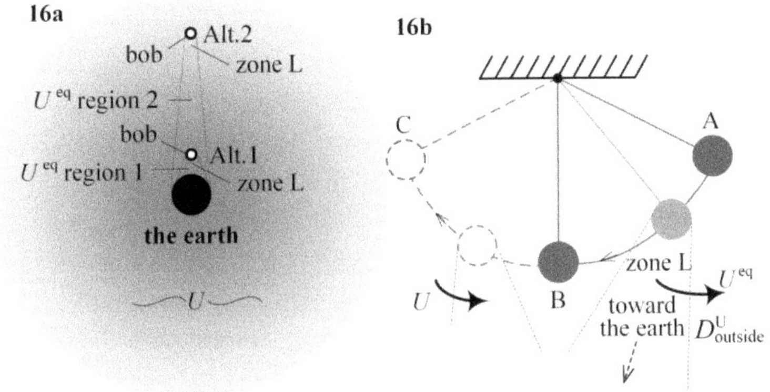

Fig. 16a Pendulum clocks closer to the earth (Alt.1) and farther from the earth (Alt.2). The release of energy U is expected to be from zone L where bonding of energy U^{eq} is the weakest.
Fig. 16b The speed of the release of energy U^{eq} from zone L increases with distance from the earth as the density of energy U adjacent to that zone ($D^{U}_{outside}$) decreases. The U^{eq} region of the bob-earth system is marked by fine-dashed lines.

According to law Ia, equilibrium energy U^{eq} has to be released from the bob-earth U^{eq} region as the bob moves from A to B (Fig. 16b) owing to the spontaneous-toward-the-equilibrium work by nature. The speed of the bob along A-B pathway is restricted by the speed of the release of equilibrium energy U^{eq} from the bob-earth U^{eq} region, as indicated in relation (20). For the clock to tick faster, energy U^{eq} has to be released faster.

It is expected that energy U^{eq} is released from zone L, which is located inside the bob-earth U^{eq} region, and very close to the bob. That is for the gravity forces that bind energy U^{eq} to the earth and the bob are weakest in zone L, whence energy is expected to be released from zone L obeying the least-effort rule (section 34).

The speed with which equilibrium energy U^{eq} is released from zone L must depend on the density of energy U outside the bob-earth U^{eq} region (denoted D^{U}_{outside}) in the area surrounding zone L. The lower the density D^{U}_{outside}, the faster release of energy U^{eq} is expected. As the clock is lifted toward a higher altitude (e.g., from Alt.1 to Alt.2 in Fig. 16a), density D^{U}_{outside} in the space adjacent to zone L decreases. Thus, it is expected that the clock will tick faster at a higher altitude, where the speed of the release of energy U^{eq} is faster. That would explain the gravitational time dilation.

The above explanation doesn't involve time that is "flowing" faster or slower. Time doesn't flow. Time just is. It is the bob that moves faster or slower owing to material energy U that flows out of and into U^{eq} region slower or faster. That applies to every other process (chemical, biological, etc.) that involves motion.

The release of energy U^{eq} due to bob's motion along pathway A-B is spontaneous owing to the execution of the SLT eradicating nonequilibrium in the bob-earth system. The absorption of energy along pathway B-C is forced by nature's contra work that brings the bob-earth system toward a lower degree of equilibrium. As the differences in the degrees of equilibrium between A and B, and between C and B are the same at any given altitude, the amounts of energy released (between A and B) and absorbed (between B and C) have to be equal. The lengths of pathways A-B and B-C are equal as well. It follows that $t^{\text{A-B}} = t^{\text{B-C}}$. In a full cycle of bob's motion, there is no net change in the equilibrium energy contained in the clock-earth system. The spontaneous work by nature (done from A to B) equals the contra work (done from B to C) such that the relation (21) is satisfied.

The pendulum clock scenario demonstrates that the motion of energy U (release/absorption) can be accelerating/decelerating. This implication could be seen controversial as the speed of the photon (equal to the speed of the motion of energy U, as will be discussed in section 44) is normally assumed to be a constant of nature. According to FLH, however, the speed of the photon cannot be an absolute constant as constants in physics are believed to be emergent properties of interactions (section 3). The speed of the photon is expected to be an emergent property of the motion of energy U_{fluid} (the same as the motion of energy U_{photon}), which results from the execution of the SLT as will be suggested in section 44.

It may be of interest to note that the release and absorption of material energy U associated with swinging of the bob can occur in packets only, released along pathway A-B, and absorbed along pathway C-B in the same-size packets.

36 Sun-planet interaction

A planet orbiting the sun is illustrated in Fig. 17. As the planet and the sun are in relative motion and are connected by G-P field, they must be subject to gravitational interaction. As nature operates in instants of space (section 4), the planet-sun's centripetal direction is the line of the interaction, but not necessarily the line of a gravity force. The sun-planet distance doesn't change because the planet's orbit is circular. (The actual ellipse-like rather than circular orbit can be ignored for the purpose of this discussion.)

As the planet is in motion, it has to be subject to the real force of inertia (F_i), whence it has to be subject to the direct force of gravity (F_g^δ) of the same strength as force F_i while acting in the opposite direction, consistent with the action-reaction law. Force F_g^δ must be generated by gravitational interaction of the planet with the sun, for we don't know of any other potential source of that force. (It surely isn't an electric force.) Force F_i is generated by the interaction of the planet in motion with distant masses. It has to act in the direction opposite to motion, *whence the direct force of gravity F_g^δ has to act in the direction of planet's motion.*

No centripetal force F_{cp} between the planet and the sun exists. Interaction in centripetal direction must be nonquantitative for the

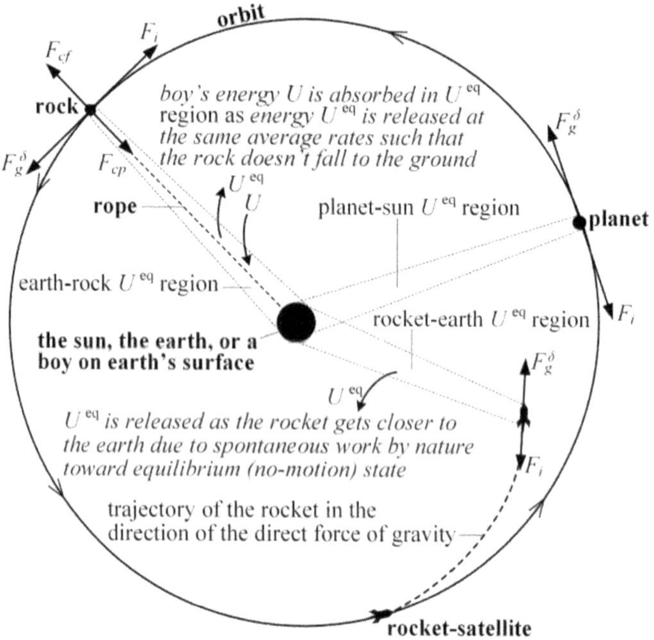

Fig. 17 A planet orbiting the sun forms a planet-sun system in meta-equilibrium as no energy U^{eq}/U is released/absorbed from/into their U^{eq} region. The degree of the planet-sun equilibrium doesn't change. Neither centripetal (F_{cp}) nor centrifugal (F_{cf}) force exists. The direct force of gravity and the force of inertia act tangential to the orbit such that the action-reaction law is obeyed.

The earth-orbiting rocket-satellite briefly slows down. With its engines off, it follows variable direction of the direct force of gravity.

A boy standing on the earth is doing cyclic work by pulling-in and letting-out a rock attached to a rope. Centripetal (by the boy) and centrifugal (by inertia) forces are real (nonfictitious) and variable, as controlled by the boy. The boy forces his own energy U into the rock-earth U^{eq} region. He loses his mass (material energy U) whence the degree of the boy-earth equilibrium increases.

planet and the sun remain at a constant distance from each other such that *the degree of internal equilibrium of the planet-sun system doesn't change*. The normal view is that a centripetal force, F_{cp}, exists and is balanced by the centrifugal, fictitious force, F_{cf}. Both those forces are supposed to act in the planet-sun centripetal direction. There appear to be two problems with the normal view. First, the inertial force—which is generated by the interaction of a planet with distant masses according to both Mach and the EE theory suggested in this book—must act in the direction opposite to planet's motion, while there is no motion of the planet in the planet-sun centripetal direction, that is, there is no inertial effect in that direction. Second, the centripetal force is considered being a (physically) real force while the inertial (centrifugal) force is considered being a fictitious force. By definition, a fictitious force exists in the physicist's mind only. A real force cannot be balanced by a force that exists in the physicist's mind only, regardless if the physicist is in accelerated motion or not. Thus, nature isn't doing any work in the centripetal direction. The planet-sun system is in a continuous state of (meta-)equilibrium as there is no motion in that direction. What that means is that no material energy U^{eq}/U is released/absorbed from/by the planet-sun U^{eq} region, consistent with law Ia (section 3).

Concerning the equilibrium of the planet-sun system, a motion of the planet and the sun toward (or away from) each other would mean a spontaneous motion from the state of (meta-)equilibrium to a state of nonequilibrium. That would contravene the SLT. If there is no gravitational wave of significance entering the planet-sun U^{eq} region (i.e., there is no contra work of significance done by nature on the system), no energy U of significance is absorbed inside the U^{eq} region and the planet isn't pushed outside the orbit. As a result, the planet-sun system remains in a state of equilibrium and the planet's orbit remains approximately circular.

The planet doesn't accelerate along its orbital path. Contrary to the normal view, it doesn't accelerate in the centripetal direction either. Also contrary to the normal view, there is no real force and no real acceleration in the centripetal direction. Those appear to be mathematical constructs. Concerning the planet-sun motion, there can be no acceleration due to spontaneous work by nature as such an acceleration results from an increase in the degree of equilibrium and the associated release of material energy U^{eq}, as discussed in section 22. In other words, *in the planet-orbiting-the-sun scenario, there is no release of energy* U^{eq} *for there is no change in the degree of equilibrium (law Ia)*. (That appears to be analogous to the chemical synthesis and dissociation reactions: no such reactions can occur without the release/absorption of heat and the associated change in the equilibrium of the substance.)

It has been suggested above that a centripetal force doesn't exist. Nonetheless, the planet and the sun actually do interact in their centripetal direction, which is a nonquantitative (nonobservable) interaction. That is consistent with FLH according to which nature operates in instants of space, which are directions in space (section 4), for instance, the planet-sun centripetal direction. But *the planet also interacts (quantitatively) with distant masses in the direction of motion*, which is the direction of planet's orbit. It is worth noting that when an apple is falling from a tree, both interactions are quantitative and occur along the same line, as explained in the next paragraph.

A rocket-satellite orbits the earth with its engines off (Fig. 17). It is moving along a circular pathway at a constant speed. The astronaut briefly fires the engines forward to slow the satellite down. As the satellite engines are off again, the trajectory of the satellite moves inwards the original orbit and follows a spiral pathway. The degree of equilibrium of the satellite-earth system increases (the satellite is getting closer to the earth), as energy U^{eq} is released from the satellite-earth U^{eq} region. As the degree of equilibrium increases, the rate of energy U^{eq} release increases, and the rocket accelerates along its pathway, in accordance with relation (19). The direction in which the direct forces of gravity (F_g^{δ}) and the inertia forces (F_i) act is always the line of the rocket motion for the action-reaction law must be obeyed.

At the time the astronaut slowed the satellite down, the force of gravity acting in the direction of the satellite's orbit was reduced. However, according to law (9), force F_g^{δ} cannot change if distance d doesn't change. Hence, it was only the component of force F_g^{δ} acting along the orbit ($F_{g(\mathrm{orbital})}^{\delta}$) that was reduced while the direction of force F_g^{δ} was changed toward the direction of the spiral. As a result, the centripetal component of F_g^{δ} ($F_{g(\mathrm{centripetal})}^{\delta}$) appeared and inflicted a conflict. A toward-the-increase-in-equilibrium motion of the satellite was initiated. Before the satellite slowed down and $F_{g(\mathrm{centripetal})}^{\delta}$ appeared, there was no conflict in the satellite-earth centripetal direction. Consequently, there was no toward-a-greater-degree-of equilibrium motion between the satellite and the earth in that direction.

A boy standing on the earth and swinging a rock attached to a rope (Fig. 17) releases his energy U by swaying his hand. The boy-earth degree of equilibrium increases as the boy releases his mass (energy U). The amount of energy U the boy releases in each swing equals the amount of energy U that is absorbed in the rock-earth U^{eq} region to become energy U^{eq} while the rock is brought up. It also matches the amount of energy U^{eq} release associated with each momentary fall of the rock. In effect, the boy is doing a bit-by-bit contra work and the rock is not falling. As the boy releases his energy, his mass decreases such that the degree of equilibrium of the boy-earth system increases. As a result, the distance between the boy and the center of the earth can remain unchanged. In terms of equilibrium, it is the boy rather than the rock is "falling" for he gets closer to equilibrium with the earth as he gets thinner. The degree of equilibrium between the rock and the earth doesn't change except for the recurring falls of the rock. The material energy U (mass) that the boy releases is miniscule as compared to the masses of the earth and the boy, yet it is crucial to explaining the boy-swinging-the-rock-on-the-rope scenario.

Let a handle be attached to the back of the planet illustrated in Fig. 17. Let superman, who is stationary in the sun's frame, grab and the handle. The planet stops moving. Superman feels a jerk of a force acting tangential to the orbit, which may be called an "orbital force." If the orbital force felt by the superman weren't the same as the direct force of gravity, what could it possibly be? There are no other known forces that could be acting here. (It indeed wasn't an electric force.) The experiment thus confirms that a gravity force acts on a planet orbiting the sun in the direction of the orbit. That force acts along the line of motion which, in any instant of time, is the direction of the force of inertia. The orbital force (force F_g^{δ}) has to act in the direction of the planet motion owing to the existence of the force of inertia generated by the planet's motion, and be in agreement with the action-reaction law. The direction of the direct force of gravity F_g^{δ} is set by the motion direction, just like in the case of a falling apple.

A cannonball experiment is illustrated in Fig. 18. Before the cannon is fired, springs A and B are unstrained. After the cannon is fired, the trajectory of the cannonball is approximately parallel to the earth surface over the initial very short distance Δx. Hence, spring A remains practically unstrained over Δx. The experiment confirms no centripetal force acting between the cannonball and

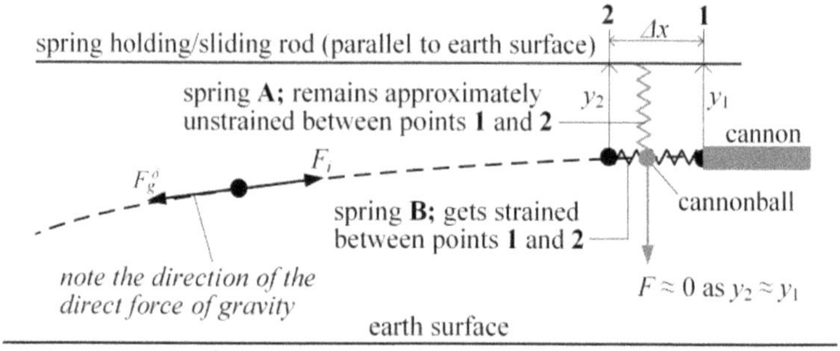

Fig. 18 Cannonball experiment reveals nonexistence of centripetal force as no strain of significance is measured in spring **A** over the initial small distance Δx.

the earth. For a centripetal force to exist, it would have to be balanced by a force of inertia, however, as there is practically no motion in the centripetal direction over distance Δx, the force of inertia in this direction is nil. The straining of spring B confirms that a force of gravity between the earth and the cannonball acts in the direction of motion, that is, opposite to the force of inertia, just as in the case of the orbital force acting on a planet orbiting the sun. As the earth and the cannonball are electrically neutral, there seems to be no other than the gravity force that could cause the straining of spring B. Both the superman and the cannonball experiments confirm that *the direct force of gravity acts in the direction of motion, which has to be opposite to the direction of the force of* inertia for the action-reaction law has to be obeyed, just like in the case of an apple falling from a tree.

Suppose a cannonball is fired at an initial speed of about 8 km/s. Spring A will remain unstrained as the cannonball travels around the earth. Nature will run a cannonball around the earth, like she runs a planet around the sun, along the least-effort trajectory. Those are the only trajectories along which no energy U^{eq}/U is released/absorbed from/by the U^{eq} regions (no work is done by nature) as the degrees of the equilibrium of the planet and the sun, and the earth and the cannonball systems don't change. That is similar to the weak-point-in-the-balloon scenario (section 34): *Nature, in each instant of time, executes her laws in all directions in exactly the same manner*, and the motion of the planet (or the breaking of the balloon) emerges in the least-effort direction (or at the thinnest balloon rubber location).

The direction of gravity force isn't set by the straight-line connecting the gravity centers of two objects appears to be a fact of everyday experience. The most familiar scenario is a car driver taking the foot off the gas pedal on a flat highway. The car slows down until it comes to a stop. It will stop because of friction. Never mind the friction, the point is that as the car is still in motion, there is a force acting upon it in the direction of motion. That force is real for it can be detected and measured by attaching a spring to the rear bumper. If it isn't a gravity force, what could it possibly be? Newton and other physicists often called that force an inertia force, a force that is an innate property of a mass object. However, if the inertial effect is the result of the interaction of an object with distant masses (Mach's idea), in that case, the inertia force acting on the car cannot be an innate property of the car's mass. Moreover, the action-reaction law must be obeyed (no exceptions).

The suggested conclusion is that the force acting on the car in the direction of motion is the horizontal component of the direct force of gravity. Its magnitude decreases as the car slows down and the direction of the direct force of gravity approaches vertical. The horizontal component of the gravity force must always be balanced by the force of inertia acting in the opposite direction as required by the action-reaction law. The gravity-based-contact force balances the vertical component of the gravity force between the tires and the earth. When the car comes to a stop, the horizontal component disappears. Only then will the direct force of gravity (and not just its component) point toward the earth's gravity center. If the force of inertia is real indeed, a component of the direct force of gravity must be always in the direction of motion to obey the action-reaction law. There must be no preconceived direction of a force for it is an emergent property of interaction.

37 Newton’s first law of motion

As pointed out in section 3, Newton’s third law of motion (law VI) is suggested to be a true law of nature (an FLP), while neither Newton’s first law (law IV) nor Newton's second law (law V) can hold that status. That is consistent with FLH. Law IV presents a relation between the motion of an object and the force acting upon it. As nature is unable to realize motion (she owns no clocks and rulers), law IV cannot be an FLP. It is an emergent law discovered by the physicist. Law V, in turn, is a physical law for indirect gravitational interactions (section 18). Hence, it exists in the physicist's mind only and, as such, it cannot be an FLP either.

Law IV states that an object remains at rest or in uniform motion unless acted upon by an unbalanced force.[o] The problem with that statement of Newton's first law is that no unbalanced forces can exist according to the action-reaction law. If an object is in motion through G-P field acted upon by an "unbalanced" force, that force has to be in fact balanced by the opposite force (of inertia) of the same magnitude and acting in the opposite direction. It doesn't matter if the object is in uniform or accelerated motion.

That points to a remarkable beauty of nature. As she operates in instants of time, having the forces of inertia balance exactly other forces acting upon objects in motion ensures that the universe is in perfect equilibrium in each instant of time. Single-mass objects in motion in gravitational-potential-free space (G-P-FS) also are in equilibrium for there are no forces acting upon them. If there are objects in G-P-FS that interact and thus exert forces on each other, those forces also are balanced as discussed in the bullet-rifle scenario (the last paragraph of section 17).

Let two objects having equal masses and moving along parallel lines collide with a brick wall. Immediately prior to the collisions, the objects moved with the same speed. However, object 1 was in uniform motion while object 2 accelerated in the frame of the wall. The damage to the wall was identical in both cases for the forces acting on the objects, and the momenta of the objects were the same just prior to the collisions. However, law IV states that prior to the collisions there was no unbalanced force acting on object 1, while law V states that an unbalanced force was acting on object 2. The physicist can easily distinguish between the two different scenarios by measuring the speed of the objects prior to collisions. For nature to realize the difference in the motions of the objects, she would have to own a clock and a ruler, and do some math.

---

[o] The first (Newton's) law of motion was likely discovered by Aristotle: "Further, no one could say why a thing once set in motion should stop anywhere; for why should it stop here rather than here? So that a thing will either be at rest or must be moved ad infinitum, unless something more powerful get in its way" (Aristotle, *Physics*, BK4 §8).

That is forbidden by FLH. Therefore, law IV, as stated, is in disagreement with the FLH corollary for in executing that law, nature is permitted to do something she cannot possibly do. It appears that a more appropriate statement of Newton's first law of motion would be: *"A single-mass object will remain at rest or in uniform motion if the state of equilibrium between that object and any other object doesn't change, that is, no equilibrium energy is released/absorbed in conjunction with the motion of the object."*

The proposed statement of law IV makes it directly applicable to a spinning top and a planet orbiting the sun (section 36), just like Newton thought about it when he proposed his first law of motion.[58] In neither of those scenarios, the degree of equilibrium changes (neglect frictional effects). And, no equilibrium energy is released in conjunction with the motion in sun's frame (a planet) or earth's frame (a top). In neither case an unbalanced force exists, if one believes in the real force of inertia. Believing in the real (i.e., nonfictitious) force of inertia generated by the interaction of an accelerated object with distant masses can be hard because it means that the speed of gravitational interactions is infinite.

38 The structure of plasma star

Astrophysical research has shown that most visible stars comprise balls of plasma, which consist of free electrons and positively charged ions with relatively few neutral atoms present. Based on the prior discussions, that image of a plasma star is incomplete. A star must be filled with material energy U for the entire universe is filled with it (section 22).

The suggested structure of a plasma star filled with electrons, positively charged ions, and energy U_{fluid} still is incomplete. The visible part of star, which represents the core of a star (roughly, the plasma ball delineated by the photosphere), is a an object to which energy U_{fluid} is attached by gravity as illustrated in Figs. 8 and 19. The attached energy forms star's "atmosphere." It means that a star comprises two distinct parts: the plasma ball extending to its photosphere, herein referred to as the core of a star, and its atmosphere comprising primarily energy U_{fluid} pulled by and

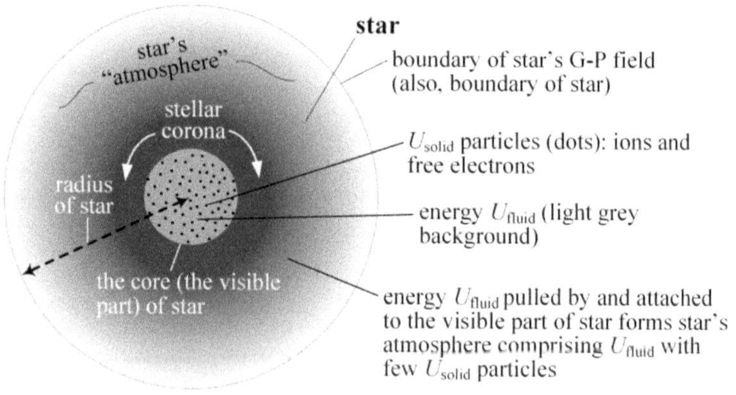

Fig. 19 Schematic structure of plasma star.

attached to the core, which extends from the surface of the core to the boundary of the G-P field of a plasma star. Thus, the rational answer to the question "how big is a star?" appears to be: "A star is as big as its G-P field."

39 Coronal heating problem

An explanation of the coronal heating mystery has been sought for over sixty years, with no success (J. A. Klimchuk[59]). A review of the most viable explanations of that mystery has been provided by M. Aschwanden[60]. The prevalent, potential explanations of the stellar corona mystery include heating coronae with the energy released by the collapse of magnetohydrodynamic waves, or the energy released from the generation of nanoflares in the magnetic reconnection process occurring within the lowest portion of a corona. Those potential explanations are intended to identify a process by which stellar coronae are heated, while obeying the SLT. The approach to explaining the stellar coronae mystery taken herein is to answer the question "*why are stellar coronae so hot*?" rather than "*how are stellar coronae heated*?"

Over a short-period of time, for instance, over a million years, a plasma star is in a near-perfect equilibrium state with respect to the distribution of its material energy U. While photon emissions, solar wind outflow, flares, coronal loops, etc., are occurring, those are very local and short-lasting dynamic phenomena that don't appreciably affect the large-scale, near-equilibrium state of the distribution of energy U inside the entire star.

A tube of material energy U cut-out in the direction of the radius of a plasma star is illustrated in Fig. 20. The tube comprises energy $U_{\text{fluid}} + U_{\text{solid}}$. As energy U distribution is in a near-equilibrium state, the SLT is barely executed. Thus, the pressure of U, which is determined by the density of U (D^U), has to be approximately equal on both sides of any cross-section located along the energy tube. That also applies to cross-section A located adjacent to the photosphere. The average density D^U just below cross-section A (D^U_{inside}) has to be approximately equal to the average density D^U just above that cross-section (D^U_{outside}). That is possible because the high densities of plasma particles (ions and free electrons) contained primarily inside the visible part of star (particle density is estimated at about 1×10^3 kg·m^{-3}) contribute very significantly to the average density D^U_{inside}. That contribution , however, quickly diminishes with outward distance from the photosphere as particle density in the corona becomes orders of magnitude less (estimated at roughly 1×10^{-13} kg·m^{-3}, and lower). Because $D^U_{\text{inside}} \approx D^U_{\text{outside}}$ at cross-section A, it must be that $D^{U_{\text{fluid}}}_{\text{inside}} \ll D^{U_{\text{fluid}}}_{\text{outside}}$ as $D^{\text{particles}}_{\text{inside}} \gg D^{\text{particles}}_{\text{outside}}$. The density of energy U_{solid} particles (the density of the cores of the electrons, the protons, and the neutrons) is expected to be orders of magnitude higher than the density of energy U_{fluid}

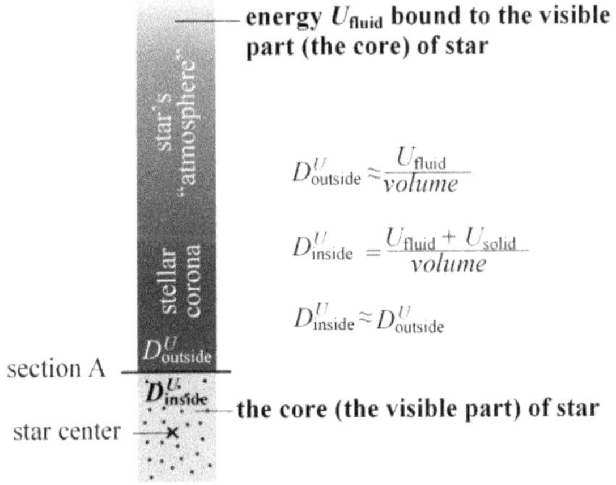

Fig. 20 Stellar corona has to exist as the densities of energy U have to be approximately equal immediately below and above section A.

that contains little or no U_{solid} particles. Therefore, as the density of plasma particles comprising U_{solid} is much lower in the corona than inside the core of the star, density $D^{U_{\text{fluid}}}$ has to be much higher in the corona than inside the core for the star to exist in near-equilibrium state.

As $D_{\text{inside}}^{U_{\text{fluid}}} \ll D_{\text{outside}}^{U_{\text{fluid}}}$, the temperature and the average vibration frequency of particles in the corona are expected to be orders of magnitude higher that those inside the core of the star, consistent with relation (25). As a result, the energies of the photons emitted from the corona (the X-ray range) are expected to be orders of magnitude higher than the energies of the photons emitted from the star's photosphere (the visible-light range).

If energy U exists as a material substance indeed, the proposed explanation of the star corona mystery appears to be highly viable. For this scenario, it is suggested that a stellar corona isn't heated as normally suggested. It was formed as the star was formed to its present size by accreting a "halo" of high pressure (high density) energy U such that the star could be stable all through the accretion process. A stellar corona provides the counterweight necessary to prevent the SLT from dissipated away a plasma star. A plasma star couldn't exist without a corona that comprises material energy U. The proposed explanation is in agreement with the fundamental characteristic of stellar coronae: the density of plasma particles is expected to decrease very rapidly over the transition zone. As a result, the $D^{U_{\text{fluid}}}$ and the temperature are expected to increase very rapidly over that zone, which is in agreement with observations.

Suppose that the visible part of a star is cooling down with time as the plasma particles are forming a continuous molecular lattice, which means that the dots (ions and electrons) shown in Figs. 19 and 20 become lattice-bonded. After the lattice is formed, the plasma particles no longer contribute to the energy pressure acting on the corona at the photosphere. The effective (i.e., contributing to the pressure) density of energy U inside the star is dropping dramatically. As the star is cooling down, the SLT removes energy U from the stellar corona to keep the densities of energy U at equilibrium at all times at all cross-sections along the tube. The ("over-blown") corona is disappearing. It follows that the most outstanding difference between the sun and the earth is that the visible part of the sun (the sun's core) comprises unbound plasma particles, while the visible part of the earth (the earth's core) comprises a molecular lattice, that is, lattice bounded particles. As a result, the earth has no over-blown corona.

The proposed explanation of the stellar corona mystery appears to be fairly simple. Yet, it may be seen complicated for it involves a number of new ideas that need to be put together: - the range of gravity is finite such that a star has its well-defined atmosphere orders of magnitude larger than the visible part of a star; - material energy U (matter) exist is subject to gravitational interactions with objects such as electrons, stars, etc.,; - energy U is filling the entire universe, including the interiors of stars and atoms; - the motion of star's energy U in radial direction is extraordinarily slow over a short period of time—say, over a few million years—such that the radial tube shown in Fig. 20 can remain in a near-equilibrium state; - the energy of photons emitted from the corona has to be extremely high because the density of U_{fluid} inside the corona is extremely high; - the extremely high density $D^{U_{\text{fluid}}}$ is required to keep the entire star stable with respect to the inside and the outside pressures of plasma; - high $D^{U_{\text{fluid}}}$ is associated with high vibration frequencies of the ionic electrons and the ions floating in star's corona; - high vibration frequencies result in the emission of high energy photons, as captured in the Planck-Einstein relation, $E_{ph} = hf$.

40 Quantum entanglement

To explain the quantum entanglement effect, it is first necessary to explain the idea of an "entanglement." Let a rocket be launched from the earth. After the G-P field of the rocket separates from the G-P field of the earth, the motion has to be conserved between the two G-P fields consistent with the motion-conservation law. That is for no motion has either entered or left either the combined or the separated G-P fields during or after separation. In that sense, the G-P fields of the rocket and the earth have become entangled upon separation: As motion has to be conserved, a change in the amount of motion inside one of the two G-P fields has to result in instantaneous and opposite change in the amount of motion inside the other G-P field. Once an entangled state is defined as discussed above, the quantum entanglement effect can be explained.

According to QT, for two particles to become entangled, they have to first interact and/or be spatially close. Those requirements can now be exactly specified: "Interact" means *direct interaction of particles* (i.e., the interaction of particles inside their combined G-P field), and "spatially close" means *G-P fields of the particles are connected*, which both reflect the same circumstance.

It follows that if two photons form a combined G-P field, they will become entangled after their G-P fields separate. Since, after the separation, an act of measurement made on photon 1 changes its spin, the spin of photon 2 has to be near-instantaneously subject to the same while opposite change. "Near-instantaneously" means that while the information on the act of measurement of the spin of photon 1 reaches instantaneously the vicinity of photon 2 (no matter the separation distance), the change of the spin itself isn't expected to be exactly instantaneous due to the inertial effect on a local (i.e., G-P field of a photon) scale.

With reference to the much-talked-about Bohr-Einstein debate, Copenhagen interpretation implies that a well-defined spin of a particle doesn't exist until a spin measurement is made. According to the FLH corollary, that can't be a correct implication. As nature is unable to realize time, she is unable to realize spin. Therefore, in the quantum *description* of physical phenomena, nature must be forbidden to realize (forbidden to know) spin.

As far as nature is concerned, at the time of measurement of the spin of photon 2, both photons are still inside their combined-pre-separation G-P field. The separation of G-P fields, the passage of time and the travelled distance cannot result in a violation of the motion-conservation law, for that is an exact and absolute law. (Dehumanized nature is exact and absolute according to FLH, as suggested in section 2.)

Concerning the "spooky action at a distance", there appears to be nothing spooky about the interaction of photons 1 and 2 for distance doesn't exist from the perspective of nature. In summary, nature executes the motion-conservation law regardless of space and time for she doesn't know that those exist, as implied by FLH. How the motion-conservation law is executed isn't for us to know for that law, as any other law of nature, represents an unreducible principle.

I've made several references to FLH's implication concerning nature's laws being absolute and exact. That implication in itself doesn't seem to be extraordinary. Yet, the consequences of it may be thought bizarre as the implications of quantum entanglement are examined in more detail. It is therefore worthwhile to convert the "bizarre" into a "rational."

Trillions of cosmic objects contained inside a G-P-FH form a supercluster (see Fig. 5). Let there be two separated-by-a-distance photons inside the supercluster, which are entangled. As their G-P fields, G-P field 1 and G-P field 2, are separated, the photons cannot interact directly. However, they can interact indirectly as a spin measurement on photon 1 has to inflict a conflict affecting the interaction of that photon with distant masses, which include photon 2. Indirect interaction means the interaction of the two photons via gravitational interactions with trillions of stars in the supercluster. As the speed of gravitational interaction is infinite (sections 11 and 12), gravitational information on the amount of conflict—that is, on the amount of spin change—arrives instantly at G-P field 2, after which the spin of photon 2 is changed by the amount exactly opposite to the amount of the change in the spin of photon 1. That is for the motion-conservation law is enforced. It may appear bizarre that the information on the amount of change of the spin is absolutely unaffected after it has interacted with the trillions of objects inside the supercluster. Yet, that is a strictly rational proposition for it stems from the following implication of FLH: Nature operates in instants of time. In an instant of time, there is no time for a spin-change information to be affected by an interaction with any of the trillions of supercluster objects. *No star and no electron in the supercluster would have time, in an instant of time, to change (affect) a spin information.*

The information on the amount of the spin change of photon 1, broken into trillions of bits by interactions with the supercluster objects, converges unaffected at the location of photon 2. (It is reminiscent of a wave function collapse.) That may be thought bizarre for the trillions of bits of information converge at the exact location of G-P field 2, say, a couple of light years away from G-P field 1. Yet, this scenario is strictly rational as well: As discussed before, nature cannot possibly know that the two G-P fields are separated for she doesn't have a concept of space. (Recall that she is dehumanized.) As far as nature is concerned, the two photons are still inside their combined G-P field, in which the motion-conservation law is enforced. Nature has no knowledge of the fact that the two photons were entangled. From her perspective, the information on the amount of spin change *has never been broken into trillions of pieces*. Also, it *has never travelled through the vast distances of the supercluster*. The breaking and the travel are realized in the physicist's mind only for it is her/his mind that owns the concepts of space and time. From the perspective of nature, the information on the amount of spin change couldn't possibly be broken into bits for she has no concepts of either the information or the bits. She just executes the motion-conservation law being a derivative of the material energy-conservation law, as discussed in section 26. How does she do it, it's not for us to know. An irreducible principle cannot be explained. It's as simple as that.

41 The double-slit experiment

The mystery of double-slit experiment has been widely publicised by Feynman[61] and other physicists. The conceptual experiment set up may include an electron gun, electrically-neutral plate with two parallel slits A and B, and a screen. Included is a door that can be used to close slit A. With the door open, hundreds of electrons are emitted from the gun, one at a time, and the familiar interference pattern appears on the screen.

With door A closed, electrons pass through slit B only and no interference pattern appears. As the electrons are emitted one at a time, an electron cannot interfere with other electrons. *This part of the experiment proves, iron-clad, that something had to pass through slit A when door A was open* as an electron was passing through slit B. That something had to be somehow linked to the passage of an electron through slit B. That is for, if nothing passes through slit A (door A is closed) when electrons pass through slit B, there is no interference pattern. Closed door A means one thing only: Nothing passes through slit A. The interference pattern must be the result of door A being open. In summary, the experiment proves, beyond doubt, that the observed interference pattern results from "something" that passes through slit A as an electron passes through slit B. Said another way, if nothing passes through slit A with door A open as electrons pass through slit B, it would make no difference to the interference pattern if door A is closed or open, which would disagree with the results of the experiment.

That "something" has to exist in physical sense and be material such that it can affect the execution of an FLP. (After an electron is emitted, there are only FLPs and matter existing between the gun and the screen.) That is for the "something" results in the generation of an observable effect—which is the interference pattern—while the generation of an observable effect requires that a material object is subject to the execution of an FLP.

It appears that material energy U detected in LIGO experiment and chemical reactions is the only physically-real substance that can be identified with that "something," simply, because there is nothing else out there in the vicinity of door A, as far as we can tell. Better said, if material energy U doesn't exist in the form of waves passing through the slits (i.e., nothing passes through slit A while an electron passes through slit B), the interference pattern has no right to exist. Just like LIGO arms didn't have right to move if material energy U didn't pass through the LIGO experiment site. Let me elaborate.

An electron emitted into the space that is filled with material energy U must continuously generate waves comprising energy U for as long as it is in motion. Those are, in essence, gravitational waves that comprise U_{fluid}. As an electron and energy U in the form of waves pass through slit B, waves of energy U have to pass through slit A if the latter slit is located sufficiently close to slit B. (If the two slits are located far apart, no waves of energy U of significance would pass through slit A and no interference pattern would be expected to appear.) The electron becomes a part of the system of gravitational waves that are formed at slit B. Another system of waves is formed at slit A. Thus, with both slits open, the motion of energy U between the slit plate and the screen is expected to be in the form of two systems of interfering waves, just like water that passes two slits in a plate that causes the interference of water waves. The electron is a part of one of the two systems of waves whence, after arriving at the screen, it reveals an interference pattern. Of interest is that the interference pattern would appear even if the electrons were to be emitted at, say, two-year intervals.

Placing a photon emitting source (an observer) between the slit plate and the screen would disturb the pattern of energy U waves between the plate and the screen. That is for the motion of emitted light (U_{photon}) would rather randomly disturb the motion of energy U_{fluid} contained between the plate and the screen, thus disturbing the pattern of interfering waves. No interference pattern would be expected to appear in the presence of light emission if the source of light is sufficiently strong. Pondering the water-wave version of the experiment, that disturbance would be rather analogous to the disturbance caused by the opening of a valve on a water hose placed between the plate and the screen. The water-waves pattern between the plate and the screen would be randomly disturbed if the pressure in the water hose was sufficiently high.

While the energy U waves generated in double-slit experiments have to be so absurdly miniscule such that, most likely, we won't be able to detect such waves directly any time soon, those waves result in a clearly pronounced observable effect, which is the interference pattern. The suggested explanation of the results of the double-slit experiment, if about correct, supports the existence of material energy U. Any double-slit experiment supplements the results of the LIGO experiment in the sense that material energy U has been detected, indirectly, in both experiments.

CHAPTER VIII – THE SPEED OF THE PHOTON

42 The speed of light

While the premise of the constancy of the speed of light has been widely accepted, some physicists have proposed that the speed of light may actually vary with time, or location in space in relation to gravitational potential. After the special theory of relativity had been published, Einstein suggested that the idea of the constancy of the speed of light applies to "... spacetime regions of constant gravitational potential".[62] Dicke proposed that the speed of light decreases close to massive objects.[63] That proposal allowed for replacing curved spacetime with a flat space. J. Moffat,[64] and A. Albrecht and J. Magueijo[65] proposed that in the early universe the speed of light was superluminal and decreasing. That proposal was intended to resolve the initial (i.e., shortly after the big bang) values problem. In general, the proposals concerning varied speed of light concentrated on mathematical assumptions. The physics underlying the constancy of the speed of light—most notably, the existence of an aether—has been given little attention since 1905. Prior to that time, intensive studies of hypothetical luminiferous aether were carried out by Augustin Fresnel, George Stokes, Albert Michelson, Hendrik Lorentz, Henri Poincaré, and others. Conducting such studies has not been entirely abandoned in more recent times (P. Dirac,[66] J-P. Vigier,[67] J. Levy,[68] M. C. Duffy[69].) Those studies and proposals appear to suggest that the principle of the constancy of the speed of light and the notion of a luminiferous aether, both famed 19th century puzzles, haven't been satisfactorily addressed so far. In particular, they suggest that the speed of light c might not be a constant built into the fabric of nature.

In the following, the unique characteristics of the speed of the photon are discussed. It was suggested in section 24 that light was a material particle comprising a quantum of energy U, called the photon, rather than an (immaterial) wave. That is in full agreement with Feynman's view.[70] Thus, when the phrase the "speed of light" is used, it actually means "the speed of the photon."

The goal in this chapter is to explain, based on the notions of FLH and a finite range of gravity, the physics that underlies the constancy of the speed of the photon. The unique characteristics of the speed of the photon include two experimentally confirmed suppositions of the 19th century. First, the speed of the photon is constant and the same for all photons being in motion *through the G-P field.* Second, the speed of the photon *in the G-P field* is independent of the speed of the photon emitting source in any observer's frame. Those two characteristics are typically referred to as "the principle of the constancy of the speed of light." As emphasized above, it is suggested herein that the principle of the constancy of the speed of light applies to the G-P field (defined in section 8), which means that it isn't universal. That is contrary to the common belief, according to which the speed of the photon (c), referred to as "the speed of light," is a constant of nature that exists, exactly the same, at each point in space and time.

According to FLH, the speed of the photon cannot be a constant of nature (section 3). As any other constant in physics, the speed of the photon needs to be considered a property of interactions that *emerges* from nature executing her laws. (It is the laws of nature that are built into the fabric of nature, and not physics constants. Physics constants are emergent physical quantities.) One of the problems with the speed of the photon being a quantity built into the fabric of nature is that it would require each photon to carry a speedometer and a speed-control device. That would contravene FLH. FLH would also be contravened by supposing that it is nature herself who somehow controls the speed of the photon. To do so, nature would have to be humanized (i.e., to have a memory, a machinery to do things on demand, ability to do mathematics, remember things, etc.). On the other hand, the speed of the photon being an emergent property of one or more interactions requires no humanization of nature.

43 Maintaining the speed of the photon

Consider a region in the universe with no gravitational potential. Gravitational-potential-free region (G-P-FR) of the universe will also be referred to as gravitational-potential-free space (G-P-FS). As discussed in section 12, such a region will exist where all mass objects in the universe are located too far from that region for their combined G-P field to extend into it. Let a photon emitted in G-P field and moving with the speed of light c enter a G-P-FS. Inside a G-P-FS, the photon isn't subject to any interactions with other objects. It therefore maintains its speed at c because motion must be conserved. Let that photon collide with another photon or a particle and then bounce off it. The photon normally is assumed to continue to move with speed c also after a collision. However, a collision of the photon in a G-P-FS would in general change its speed, just like kicking a flying soccer ball would change the speed of the ball. As the photon doesn't gravitationally interact with any objects inside a G-P-FS, its motion is not affected by any interactions. It follows that to set and maintain its speed at the speed of light c after a collision in a G-P-FS, the photon would have to have a memory of the value of speed c, measure its own positions and the times between the positions, do appropriate calculations, and have a built-in speed-control mechanism. Owing to FLH, the idea of the photon having those abilities and devices has to be rejected. Therefore, the supposition that a photon after a collision with another photon or particle in a G-P-FS must move with speed c cannot be substantiated. In summary, in a G-P-FR of the universe, the photon isn't subject to any interactions with other objects, whence the speed of the photon cannot be controlled by anything except by the photon itself, which is untenable according to FLH. It follows that in a G-P-FS, the speed of the photon can be different from speed c.

Let a photon be moving with speed c inside a G-P field, for instance, inside the G-P field of the Local Group, and then collide with and bounce off another object. Before and after the collision, the photon has been subject to gravitational interactions with other objects (stars, planets, etc.), and it is a well-confirmed fact of experience that the photon would be moving with the same speed c also after the collision. The above discussion suggests that *it is the G-P field that has something to do with maintaining the speed of the photon at* c *as the photon is running through the field.*

44 Steady-state speed of the photon

Let there be an interstellar object (it could be a star), the G-P field of which isn't connected to the G-P field of any other object. The degree of nonequilibrium in the distribution of material energy U between the core of the object and the adjacent G-P field is incredibly large. The U_{solid} cores of the particles the object comprises are the primary contributors to the nonequilibrium. By executing the SLT, nature eradicates that nonequilibrium such that the energy U (matter) the object comprises is dispersed into the adjacent G-P field (Fig. 21 - short arrows). In agreement with observations, in the time framework available to a human observer the energy is primarily dispersed in the form of the photons comprising U_{photon} (see also footnote h). After their emission, the photons are in motion (Fig. 21 - long arrows) caused by nature executing another aspect of the SLT, as will be explained next.

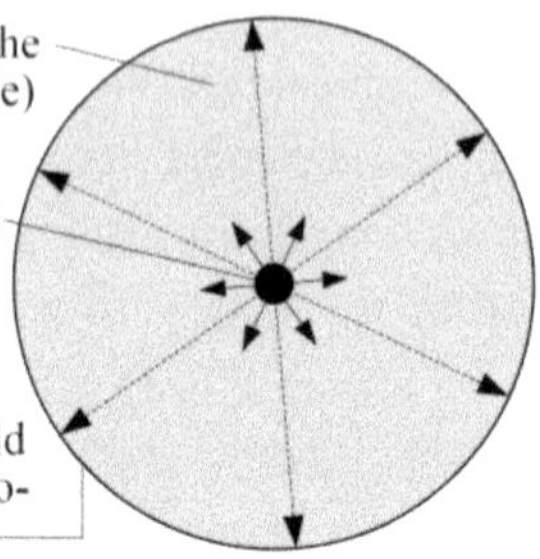

Fig. 21 Emission of the photon. The SLT is executed by nature with respect to two nonequilibrium conditions. The magnitude of gradient D in the stage 2 execution, which controls the motion and the speed of the photon, numerically equals to the average mass density D in G-P field. The magnitude of D is constant in space and time.

Any single-mass object comprises its "visible" part (the core) and energy U_{fluid} bound to it by gravitational pull, as illustrated in Fig. 8. The amount of energy nonequilibrium in a g-object, such as g-object m_{1234} illustrated in Fig. 22, can be determined from an examination of the densities of material energy U inside the g-object. Consider the portion of the G-P field of g-object m_{1234} *outside* the cores of the four single-mass objects. That portion of the G-P field has a volume of v_{outside}. Volume v_{outside} is nearly empty of U_{solid} particles, with the value of average D_{outside} being very close to zero as compared with the average energy density in the total volume of the G-P field of the g-object, denoted in v_{1234}. Volume v_{1234} is only marginally larger than volume v_{outside} while it contains relatively large amount of energy U in the form of both

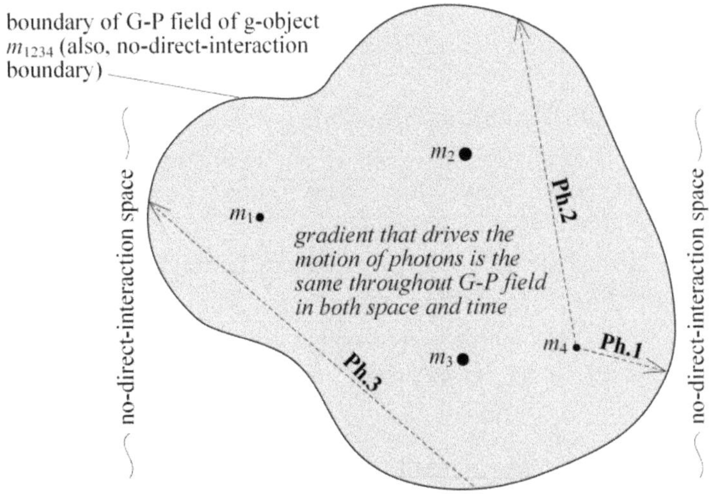

Fig. 22 Emission of photons 1 and 2 inside g-object m_{1234}. G-object m_{1234} may be the solar system or a galaxy. Photons move at steady state as the SLT is executed at a constant gradient D.

U_{solid} cores of the particles and the relatively high-density energy U_{fluid} contained inside the cores of the four single-mass objects. The average D_{outside} (applies to volume v_{outside}) can be assumed to be approximately zero when comparing with the average energy density D in G-P field (applies to volume v_{1234}),

$$D_{\text{outside}} \approx 0. \tag{27}$$

Since the single-mass objects are very small in comparison with the total volume of g-object v_{1234}, the density D_{outside} corresponds approximately to the state of g-object's *maximum nonequilibrium* with respect to the distribution of energy U. Said another way, the state of maximum nonequilibrium arises for the v_{outside} portion of the total G-P field is nearly empty of energy U.

The hypothetical state of g-object's *minimum nonequilibrium*, that is, the state of perfect equilibrium, would exist if all energy U contained inside the G-P field of a g-object were to be distributed uniformly throughout the total volume of g-object's G-P field. From postulate (1), the energy density at each location in the total volume of the G-P field, which is denoted D_{total}, would then be,

$$D_{\text{total}} = D, \tag{28}$$

where D is the average density of energy U in G-P field. Eq. (28) is correct in mathematical terms only. In physical terms, it refers to a hypothetical scenario and, in itself, doesn't seem to have any meaning.

The difference in energy equilibrium between the maximum and the minimum conditions represents a *gradient* that causes the motion of the photon. It will be called gradient D. The magnitude of gradient D is determined from Eqs. (27) and (28),

$$D = D_{\text{total(max.equilibrium)}} - D_{\text{outside(min.equilibrium)}}. \tag{29}$$

According to postulate (1), gradient D is independent of time and space, including direction. A photon present in a G-P field must always be in motion for gradient D, which has a magnitude of the average mass density in G-P field always exists. It cannot be zero.

Nature executes the SLT to bring energy U contained in the G-P field of a g-object to equilibrium. The SLT is executed by means of enforcing gradient D. The execution generates a force (denoted "force D") acting on the emitted photon. Its magnitude doesn't depend on space (location and direction) and time because it is generated in response to gradient D. Gradient D under which the SLT is executed emerges because the density of energy U is the lowest alongside the boundary of the G-P field. The photons are in motion because there is more energy U_{fluid} in the G-P field of a g-object that can be bound to the single-mass objects that form the g-object.

Thus, the photons emitted by the four single-mass objects (Fig. 22) carry excess energy U in the form of energy U_{photon}. Energy U carried by the photon can neither accumulate nor decumulate in the locality of the photon. That, in conjunction with the constant gradient D, points to a *steady-state condition*, the scenario under which the speed of the photon c is always the same. Furthermore, the loss of energy U is expected to be proportional to the distance the photon travels.

The steady-state motion of the photon described in the prior paragraph is consistent with any other thermodynamic process, in which there is no accumulation or decumulation of matter, while the driving gradient has a constant value. The steady-state motion, however, may happen only after a short period of time during which the photon accelerates to speed c.

Let a photon be emitted. There must be a short period of time needed to accelerate the photon to speed c. That is for prior to the emission of the photon, the material energy that the photon will comprise after emission isn't in motion except for some vibrations (section 33). The acceleration of energy is a direct consequence of the photon being a quantum of material energy U: No material substance can be accelerated from speed zero to speed c in an instant of time. The acceleration of the photon could be hard to accept in view of the current physics, however, it could be harder to accept that the photon is *immaterial* (see the first paragraph in section 24) such that it can accelerate to speed c instantly.

In summary, owing to the execution of the SLT by nature, the emitted photon is driven toward the boundary of the G-P field of the g-object that the photon was emitted in, for instance, the G-P field of a cluster of galaxies. The photon remains in motion due to gradient D, which has the same magnitude regardless of both the direction of motion and the distance to the boundary of the G-P field. That is consistent with FLH: Gradient D cannot depend on a distance for nature is unable to realize space.

That density D represents the gradient that drives the motion of the emitted photon can be concluded from another consideration. To arrive at that conclusion, examine a single-mass object such as the planet shown in Fig. 21. There is mass m_0 located at the center of the G-P field. In terms of gravitational interactions, effective mass along the-no-direct-interaction boundary can be assumed at zero. That is for no object inside G-P field, including the photons, can directly interact with any objects outside that boundary. Thus, the difference in the amount of mass, which is $m_0 - 0$, is spread over the volume of G-P field, v_0. The gradient of mass-energy distribution is $(m_0 - 0)/v_0 = \nabla m = D$. Gradient ∇m has a dimension of $[M{\cdot}L^{-3}]$, consistent with the concept of gravitational potential quantized by its *volumetric* amount V (section 8).

As nature is driving each photon present in a G-P field of a g-object toward a state of equilibrium, the photon–g-object system must be releasing energy U consistent with law Ia. That is similar to the release of energy U by an apple and the earth in a free fall toward each other (section 22). However, neither an apple nor the earth can release energy U themselves. During the apple-earth free fall, the energy U is released from the apple-earth U^{eq} region according to the following principle (law Ia): An increase in the equilibrium of a system requires a release of energy U. A release of energy U in the case of the photon driven toward equilibrium is subject to the same principle. That can be imagined by considering the release of energy U from the outskirts the photon, which plays the role of zone L. (For the explanation of the zone L concept see section 34, Fig. 9 and Fig. 14.) In effect, the degree of equilibrium increases, energy U is released, while the photon is subject to cosmological redshift. (The photon may be subject to a blueshift if it is approaching a single-mass object.)

The execution of the SLT as the process causing the emission and motion of the emitted photon appears to be a well-justified proposition for each object emits photons, which is what nature enforces throughout the universe bringing photon-emitting matter toward equilibrium in the distribution of energy U. That applies to a G-P field. In a G-P-FS, nature is unable to control the motion of emitted photon for there is no mediating G-P field and no gradient D.

Thus far, the discussion of the speed of the photon has been limited to the photons emitted inside the G-P field of a g-object (i.e., Ph.1 and Ph.2 in Fig. 22). The photons that may enter the G-P field of a g-object from an outside G-P field (i.e., Ph.3 in Fig. 22) have been considered. Those photons are expected to be indistinguishable from the photons emitted inside the G-P field of a g-object. That is, for they are expected to move with the speed controlled by the same, constant gradient D. In other words, the photon is expected to move with the speed of light c as long as it remains within a G-P field. The velocity of the photon inside a G-P field can change as the photon interacts with massive objects, just like the change in the velocity of the light beam passing next to the sun observed in the famous 1919 experiment. The direction of motion is an emergent property of an interaction. The photon is expected to move at the same speed in any direction. That is for gradient *D* is independent of the direction in a G-P field.

It seems that the most persuasive argument in favor of the constancy of speed of the photon being the result of the execution of the SLT, is the outcome of the experiment that has been carried out thousands of times. Let a light beam enter and exit a body of glass. The speed of the photon before entering and after exiting the body will be the same, while that speed will be lower when the beam passes through the glass. *That emphasizes that the speed of light isn't a constant of nature*. Why would those entering and after-exit speeds be the same? If the light beam, that is, each photon carries a speed-control device that can be used to control its speed, then the answer is simple: The photon remembers that it has to move outside the body of glass with speed *c*, so it measures its own speed and adjusts it as necessary to return to speed *c* upon exiting the glass. Still, that answer is even simpler if the photon is a fundamental particle in the sense of FLH—that is, it doesn't know and remember speed *c*, and it carries no speed-measuring and speed-controlling devices, which means that the photon has been dehumanized. Under the latter scenario, the speed of the photon upon exiting the body of glass is again controlled by the same gradient *D* and the same conflicts with the execution of the SLT as the gradient and the conflicts that maintained the speed of the photon at *c* before entering the body of glass. While moving inside the glass, the motion of the photon still is controlled by the same gradient *D*, but it is subject to much stronger conflicts as compared with the conflicts that affect the motion of the photon outside the glass. The stronger conflicts result from gravitational interactions of the photon with glass molecules.

The preceding discussion concerns the speed of the photon, a particle comprising a quantum of energy *U* (U_{photon}). Based on that discussion, it can be expected that also the speed of energy U_{fluid}, denoted c^{U}, released in the form of gravitational wave is the same as the speed of the photon *c*. Both the photon and the gravitational wave comprise energy U_{fluid} and both are subject to the execution of the SLT under the same gradient *D*. The density of energy adjacent to the source that releases gravitational wave has to be in nonequilibrium (i.e., elevated) in relation to the space adjacent to the source, just like in the case of the of the emission of the photon. (No material object, including gravitational wave, can accelerate from speed close to zero the speed c in an instant of time.) In other words, it is suggested that the motion of a gravitational wave is caused by the execution of the SLT, which means that nature is doing spontaneous work, which eradicates nonequilibrium in the distribution of energy U. Away from the source, it is expected that the speed of the motion of gravitational wave (c^{U}) is the same as the speed of the light (c).

In section 20, it was suggested that the principle of equivalence can be concluded directly from FLH provided, however, that the boundary conditions are the same for all free-falling objects with different masses. The boundary conditions were not specified for nature's work, which results in the release and motion of energy U, wasn't recognized at that point. As equilibrium energy U^{eq} is expected to exit U^{eq} regions of different objects in free fall toward each other with the same constant speed c^{U} that is unaffected by their masses, that speed represents a boundary condition that is the same for all different-mass objects in free-fall.

45 Independent of the motion of the observer

A moving star and two observers, Obs.1 and Obs.2, are located in the G-P field of a galaxy (Fig. 23). Both observers are stationary in the frames of their own G-P fields as well as in the G-P field of the galaxy. The star moves at speed v_{star} relative to the observers. The G-P field of the star (the dashed circle in Fig. 23) isn't moving with the star for any G-P field is absolutely motionless. Let

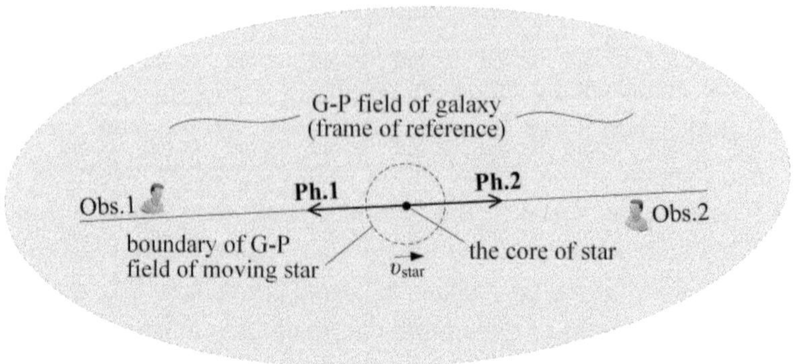

Fig. 23 Photons emitted from the core of a moving star.

photons Ph.1 and Ph.2 be emitted from the star toward the two observers. According to the prior discussion, both photons move with constant speed c in the frame of galaxy's G-P field as the magnitude of gradient *D* doesn't change with the location and it doesn't depend on the direction in the field (section 44). Because the observers are stationary in the frame of galaxy's G-P field and the photons are moving with the same speed *c* in that frame, both observers will measure the speeds of passing photons to be exactly the same regardless of the speed of the emitting star.

Suppose now that the two observers are moving in the frame of the galaxy's G-P field. The photon speeds, as measured by the observers, will be the same again regardless of their motions. That is because the G-P fields surrounding the observers and the G-P field of the galaxy represent one and the same G-P field that is motionless and stationary relative to the observers, while gradient *D* doesn't depend on the location and direction in the field.

In summary, as long as the photon remains within the G-P field in which it was emitted, for instance, in the G-P field of the Local Group, its speed will be maintained constant and independent of the speed of the photon-emitting source. It is constant because it is controlled by the directionless and constant gradient *D*, and because the G-P field is a motionless frame of reference in relation to which the speed of the photon has be measured. Maxwell's 150-year-old question, "In relation to what frame of reference is the speed of light constant?" can therefore be answered: "In relation to the G-P field introduced in section 8."

46 Michelson-Morley's experiment

Fig. 24 illustrates the result of Michelson-Morley's experiment.[71] It can be appreciated accounting for the proposed concepts of the G-P field and the speed of the photon controlled by gradient *D*, which is independent of a direction and has a constant magnitude in G-P field.

As indicated in Fig. 24, the velocities of the photons traveling back and forth in earth laboratories are always the same relative to the earth regardless of the motion of the earth in the frame of the G-P field of the solar system, Milky Way, Local Group, etc. That was demonstrated by Michelson-Morley's experiment, and is

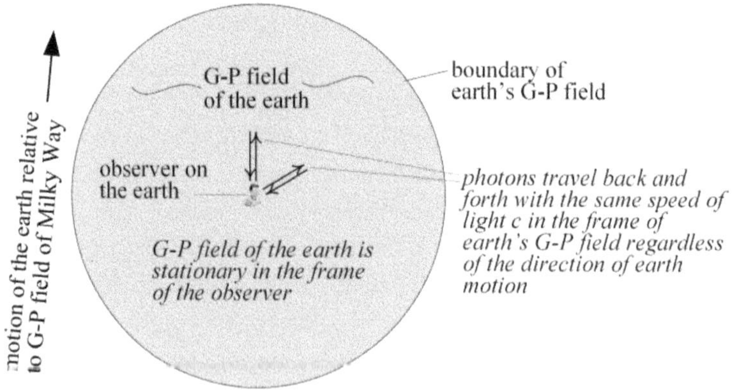

Fig. 24 Michelson-Morley's experiment. G-P field of the earth is stationary relative to the earth. In earth laboratories, the speed of light in any direction is measured to be the same relative to the earth regardless of the speed and the direction of earth's motion.

explained below. That explanation presents a restatement of the conclusions derived in the preceding sections.

In a G-P field, the photon travels with speed *c* in the frame of the field, which is stationary relative to an object, such as a planet or a star, located in that field (section 12). Except for its size, a G-P field cannot change according to postulate (1). The magnitude of gradient *D* is constant, and the same in each direction. That is consistent with dehumanized nature. She cannot do mathematics and measure things. Therefore, she wouldn't be able to tell one point of G-P field from another. The motion of the photon in the G-P field cannot possibly affect the field. And vice versa. The only thing that any object can do with respect to a G-P field is to extend and/or shrink its boundary (section 12).[p]

A fully entrained, motionless aether, herein called the G-P field, cannot affect the motion of the photon depending on the motion of a laboratory. The G-P field and a laboratory constitute a rigid body. The result of Michelson-Morley's experiment seems to be obvious if completely entrained aether is identified with (perfectly homogeneous and isotropic) G-P field. Concerning aether, the G-

[p] Reminding the meaning of the G-P field could be helpful here. The G-P field is an information field that comprises "yes" and no," which are nonquantitative signifiers. "Yes" or "no" at a location in space and time means that gravitational potential is there, or it isn't. Nature and the physicist can merely determine if G-P field is present at (yes) at or absent from (no) a location in space and time.

P field concept proposed in this book is somewhat similar to the 1845 complete aether drag concept proposed by George Stokes, who suggested that the portions of the aether close to the surfaces of the earth and other planets were stationary relative to the planets.[72] Stokes, however, didn't specify what an aether could be in a physical sense.

47 Aberration of light

In this section, the aberration of light effect is explained based on the prior suggestions of a finite range of gravity (section 1), the infinite speed of gravitational interaction (section 11), the photon comprising material energy *U* (section 24), and the gradient *D* that controls the motion of the photon and has the same magnitude at each point and in each direction in a G-P field (section 44). The proposed explanation of the aberration of light effect is based on a conceptual-level dynamic rather than a purely kinematic (the addition of velocities) consideration.

Let a photon be emitted from the dark-colored star illustrated in Fig. 25. The earth is situated at point B and is stationary in the frame of the star. The star and the earth are connected by G-P field.

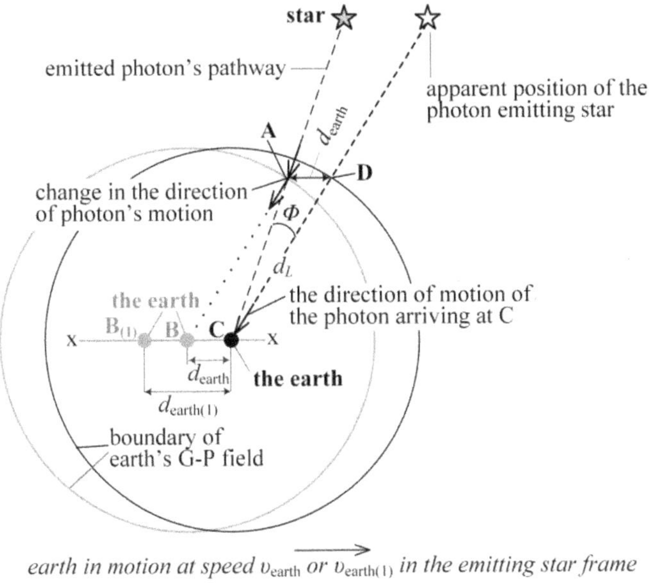

Fig. 25 Aberration of light.

An observer on the earth sees the star at its actual position. An emitted photon seen by the observer has been moving along the line marked "emitted photon's pathway." In the absence of near-by massive objects, the motion of the photon has been controlled by the execution of the SLT, which generates gradient D of the same magnitude in any direction the photon moves through a G-P field, regardless of photon's location (section 44). That direction of photon's motion has been set at the time of photon's emission.

In a more realistic scenario, the earth is moving with a uniform speed v_{earth} in star's frame along line x-x toward point C (Fig. 25). A photon emitted by the dark-colored star arrives at point A located on the boundary of the G-P field of the earth and it starts to interact directly with the earth. The earth at the arrival time is located at point B. As the photon passes point A, the direction of photon's motion starts changing for it becomes subject to direct interaction with the earth. The start of the change has to be instantaneous for the speed of gravitational interaction is infinite. The direction of photon's motion is changing toward the center of the earth.

Assumption is made that *the above change takes place instantly and in full* at the time the photon passes point A. After that change, the photon moves toward the center of the earth in a straight line until it reaches the observer on the earth. Gradient D that drives the motion of the photon is the same at any location and in any direction, which makes that assumption acceptable. Also, the medium through which the photon moves (energy U_{fluid}) is expected to have no internal friction, which further points to the acceptability of the above assumption as it indicates no frictional resistance to a change in the direction of motion. It needs to be kept in mind, however, that the above assumption presents an approximation because, in the absence of a collision, there appears to be no precedence for an instant change in the direction of momentum of an object.

Depending on the actual average mass density in the G-P field (D), the approximate time it takes the photon to move from point A to the earth ($= \Delta t_{photon}$) can be estimated from postulate (1) to be between 0.02 and 25.5 years.[q] That also is the time, denoted

---

[q] The range of possible values of D will discussed in section 63.

Δt_{earth}, that takes the earth to move with speed v_{earth} from point B to point C. The length of photon's pathway inside the G-P field of the earth is d_L (distance d_L is defined in Fig. 3).

As the photon reaches point A, the earth is at point B, in motion toward point C, moving with speed v_{earth}. Upon entering the G-P field of the earth, the photon and the earth form an interacting system in motion in the frame of the photon-emitting star. The system is formed and maintained by the direct force of gravity F_g^{δ}. There are no forces other than F_g^{δ} that could affect the direction of photon's motion in the frame of the earth (force F_D is the same in any direction). According to the assumption on the instant change in the direction of photon's motion at point A, the photon moves in the direction parallel to line A–B, which also is parallel to line D-C. It follows that the photon will arrive at the earth at point C from the direction of line D-C. The observer will experience an aberration of light effect. The aberration angle Φ is depicted by angle ABD.

Using the small-angle approximation, it follows from Fig. 25 that the aberration angle Φ by which the star appears to be shifted in view of the observer on the earth located at point C is ACD = $d_{\text{earth}}/d_L = v_{\text{earth}}/c$. The aberration angle ACD is therefore very small.

Let now the earth moving uniformly with speed $v_{\text{earth(1)}}$ be at point $B_{(1)}$ as the photon arrives at point A. Under that scenario, the speed of the earth is $v_{\text{earth(1)}} = d_{\text{earth(1)}}/\Delta t_{\text{photon}}$. Therefore, the greater is the speed of the earth, the greater is the aberration angle Φ, and the observer at point C will see the apparent position of the star shifted further to the right. The aberration angle depends on the speed of the earth (the speed of the observer) only. That speed is to be measured in the frame of the photon-emitting star.

The proposed explanation of the aberration of light effect relies on a finite range of gravity. Under that circumstance, the distance to the photon-emitting star is irrelevant as long as the individual G-P fields of the star and the earth are not connected (i.e., the star is sufficiently far away). If the range of gravity were to be infinite, the aberration of light effect would depend on the distance to the photon-emitting star.

The assumption of the instant change in the direction of photon motion at point A is in some way similar to the popular "running in the rain" explanation of the aberration of light. The direction of the rainfall corresponds to the small arrow extending from point A in Fig. 25. In view of the observer at point C, the rainfall would be tilted more in the case of the earth located at point $B_{(1)}$ than at point B at the time the photon passes point A. Granted, the speed of the earth is greater in the point $B_{(1)}$ than in the point B scenario. However, that tilt being dependent on the speed of the earth is an illusion only. The real difference between the point B and point $B_{(1)}$ scenarios is the direction of the direct force of gravity acting on the photon when hitting observer's eye at point C. The observer at point C doesn't experience the aberration of light effect because of the speed of the earth (because of the observer's speed). The observer experiences the aberration of light effect because of the direction of force F_g^δ deviates from the direction to the photon-emitting star. While physical effects can be described in terms of kinematics, those effects cannot be explained in purely kinematics terms, for nature has no concepts of space and time (section 4).

If the assumption on the instant change of the direction of photon's motion at point A is omitted, the photon's pathway inside the G-P field of the earth would be curvilinear in the frame of the earth, but an aberration of light effect would still be expected. Of interest here is that under the "instant change" assumption, any photon that the observer at point C will see will be the photon that was emitted from the star in the direction of point C. Said another way, if an observer is running, the photons the observer will see at all points of the run will be the photons emitted in the past by a star in the directions of each of the points of the run.

48 Energy and mass

The theory of relativity was formulated by Einstein[73,74,75], with important inputs from Lorentz[76] and Minkowski[77]. It is a physical law designed to describe the relations between physical events that occur in G-P field. According to FLH, the theory of relativity isn't a law of nature (nature has no concepts of time, space, or motion). Furthermore, it doesn't apply to all space and time if the range of gravity is finite. That is for in a G-P-FS, the theory of relativity doesn't apply as the speed of the light in a G-P-FS needs not to be constant.

Regardless, the hypotheses underlying the EE theory allow for explaining, in physical terms, the most outstanding mystery of the theory of relativity, which is the mass-energy equivalence stated by the famous formula $E = mc^2$. Einstein suggested a relation between mass and energy in the statement, "The mass of a body is a measure of its energy content ..."[78]. That relation is referred to as the mass-energy equivalence. It is important to remember that by "energy" Einstein meant "the ability to do work." In the EE theory, the word "energy" is used to identify either material energy U or mass m, which are one and the same as they both mean the same matter (sections 22 and 24). Therefore, the relation between energy U, which is material and physically real (nonabstract), and mass is trivial,

$$\textit{material energy } \boldsymbol{U} \textit{ and mass } \mathbf{m} \textit{ are one and the same} \tag{30}$$

While trivial, relation (30) can be useful. For instance, if energy U is forced into an object, which means that the object gets heated (section 28), the amount of mass of an object, which is the same as the amount of matter it comprises, has to increase, whence the gravitational effect of the object has to increase, all other factors being equal. Hence, the conclusion has to be that the strength of gravitational interaction of an object depends on its temperature. That is a well-known and widely accepted supposition.

The existence of material energy U indicates that the standard explanation of Einstein's equation,

$$E = m_0 c^2, \tag{31}$$

may have a physically rational interpretation. Symbols E and m_0 in Eq. (31) denote energy and rest mass, respectively. Energy E is an abstract physical entity that reflects the ability to do an amount of work. Hence, energy symbol E is replaced with symbol E^{work}, where superscript "work" accentuates "the ability to do an amount of work." Based on the discussion in section 24 and FLH, mass (material energy U) other than "rest mass" cannot exist in physical sense. Consequently, symbol m_0 can be replaced with symbol m, which represents mass (energy U). Eq. (31) becomes,

$$E^{\text{work}} = mc^2. \tag{32}$$

The term mc^2 presents the *maximum* amount of work that nature can potentially do using the entire amount of mass (m) of an object (the entire amount of material energy U the object comprises). That amount is independent of the motion of an observer. Nature can *potentially* do that work by executing the SLT in a-toward-an-equilibrium process. The maximum amount of the work, equal to E^{work} in Eq. (32), corresponds to the scenario in which *all* U_{solid} cores of the particles an object comprises are subject to a phase transition, which means they are converted into material energy U_{fluid}. As a result of the phase transition, a relatively enormous amount of energy U_{fluid} (heat) is produced. *All* energy U that an object comprised prior to the transition, including the energy U_{fluid} originally bound to U_{solid} particles, would be released under that "maximum" scenario in the form of a gravitational wave, which would move with the speed of light, in agreement with the notion discussed in section 44.

Concerning the amount of work that nature can do utilizing the motion of an object m that moves with speed v in the absence of a phase transition, that amount of work depends on the motion of an observer. The ½ factor has to be included to determine the ability to do an amount of work using the formula for the KE of object m, $KE = ½mv^2$. The other half of the total KE contained in a system is carried by the objects with which the single object m is quantitatively interacting. The KE and the mass-energy-equivalence formulae can be combined to determine the total ability to do an amount of work utilizing an object from $E^{\text{work}} = mc^2 + ½mv^2$. In summary, it is suggested that the KE and the mass-energy equivalence formulae represent the same physical law. The ½ factor isn't included in the mass-energy equivalence formula as the toward-an-equilibrium work by nature utilizes a single-mass object only. (The principles underlying the two types of work by nature as similar to the principles based on which the steam and the water fall engines are working.)

49 SR time dilation

The ticking of a pendulum clock is faster as the density of material energy U surrounding the clock is lower (section 35). That effect corresponds to the gravitational time dilation predicted by GR. Also, the lower is the density of material energy U at a location in space, the lower is the temperature (section 28). That means that a time dilation between two clocks is expected to be caused by a difference in the temperatures of space adjacent to the clocks. Say, a husband spends his life in the cold room, while his wife stays in the kitchen. As the couple meets after a long time, the husband might say, “it’s Tuesday,” while the wife might respond, “you are wrong, it’s still Monday". As suggested in section 35, the same could happen to a husband who spends his life on Altiplano while his wife stays in Antofagasta.

SR ("velocity") time dilation is discussed in this section. It is suggested that at the fundamental level, the SR time dilation is no different from the gravitational time dilation. In brief, the lower is the density of energy U adjacent to a clock, the clock is expected to tick faster. And vice versa.

First, a trivial perhaps fact needs to be emphasized: time doesn't flow slower or faster. Time doesn't flow. Time just is. And, it isn't something in itself. Time is a property of quantitative interaction. Time without quantitative interaction doesn't exist. It is defined in the human mind by the measurements of clock's ticks or frequency moon phases. Therefore, it isn't that time flows slower or faster. It is that a clock, whether it is a pendulum of a cesium clock, ticks slower or faster.

Second, the widely-known "twin paradox" cannot be explained in term of kinematics only. Pure kinematics with its components of time, displacement, velocity and acceleration, while necessary for doing descriptive physics, exists in the physicist mind only. It cannot explain the slow aging of a travelling twin. To explain that, including a dynamic effect is necessary. It is not that a travelling twin lives shorter life. It is that the U_{solid} particles that the twin comprises tick slower. That ticking is caused and controlled by forces (interactions). It will be convenient and safe to assume that the twins comprise atomic clocks.

The previously drawn conclusions need to be remembered:

- No material substance (energy U) can be accelerated from speed zero to speed c in an instant of time. That follows from the other law of gravitation, which was originally proposed by Newton as his second law of motion (section 18).
- As a material object is in motion toward an equilibrium (a no-motion) state, energy U is released (law Ia). The larger or smaller is the density of energy U surrounding the object, the energy is released slower or faster, respectively (section 22).
- Each object in motion through G-P field moves toward a no-motion state, that is, toward a G-P-FR (see Fig. 5). Thus, each object in motion through G-P field releases energy U.

Concerning SR (velocity) time dilation, the normal view is that the frequency of clock ticks (f^{ct}) is in a reverse proportion to the velocity of the clock (v), $f^{ct} \propto 1/v$. That cannot be a law of nature as nature doesn't realize velocity (or speed).

Suppose a twin left the earth in a rocket. The earth is about at rest in the frame of G-P field. The twin travelling through space (say, through the supercluster of Fig. 5) has switched the engine off and travels at a substantial speed in the same frame. Direction of the travel is irrelevant. The twin-rocket object, which comprises atomic clocks, must be releasing energy U as it moves toward an equilibrium (a no-motion) state. The energy released in the form of a gravitational wave will spread out throughout G-P field at speed c. However, before that happens, the released energy has to be accelerated to c. Owing to the acceleration, the density of U adjacent to the boundary of the object (shown in Fig. 8) increases. The increased density of U results in the frequency of the ticks of the clocks the twin comprises being lower. As the (twin-rocket) object travels faster, both the rate of energy release and the energy density increase such that the ticking of the twin's clocks slows down. The twin ages slower. In essence, the SR time dilation is the same effect as the gravitational time dilation. It's all about the release of material energy that cannot be accelerated to speed c instantly. The speed of the motion of U being released controls the speed of clock's ticks, similar to controlling the speed of an apple falling toward the earth or a feather falling toward the moon.

CHAPTER IX – EVIDENCE IN SUPPORT OF THE EE THEORY

In this chapter, comments are made concerning the viability of the proposed EE theory. Those are made with reference to the current paradigm of physics (CPP).

50 The law of gravitation

All evidence for the viability of Newton's theory of gravitation also applies to the viability of the proposed physical law for direct gravitational interactions (9). Law (9) is expected to be more accurate than Newton's law (2) because it incorporates function K. Assumption $K = 1$, which reduces proposed law (9) to Newton's law (2), makes law (9) as accurate under strong gravitational interactions as Newton's law. The proposed law (9) appears to be as viable as the Newton law.

Note that there appears to be no available data, however, from which the accuracy of laws (9) and (2) under weak gravitational interactions could be determined, for instance, the interactions of the sun with a small ice-planet situated at the outskirts of Kuiper belt.

51 Comprehensiveness

Pondering comprehensiveness, the proposed EE theory answers numerous questions. Perhaps the most striking are: Why objects are in motion in spatially-infinite universe? What is the real force that causes the earth to bulge at the equator? What is the frame relative to which Newton's bucket spins? Why a car driver feels the inertia force immediately after s/he pushes the gas pedal? Where is PE located? What is the direction of the direct force of gravity? How an object in free fall obeys the action-reaction law? Why did the hammer dropped by Commander Scott fell to the moon surface so slowly? What do gravitational waves consist of in a physical sense? What is the physics underlying the principle of equivalence? Why doesn't an atomic electron fall onto nucleus? What is the meaning of heat and temperature? What is the true size of a star? Why isn't the speed of the photon a constant of nature? What underlies gravitational time dilation? Why a stellar coronae must exist? What is the matter indirectly detected in both double-slit and LIGO experiments? Why isn't QT a weird theory contrary to the common belief? These questions don't seem to have physically substantiated answers in the CPP. The answers to those questions suggested in this book appear to provide strong, while typically indirect evidence for the viability of the EE theory.

52 Direct evidence

Suppose a spaceship carrying Newton's bucket on-board is sent to a gravitational-potential-free region (G-P-FR) somewhere in the universe where the bucket is made to spin. The water surface remains flat. That would become a direct evidence that some key gravitation-related notions of the EE theory are correct as it would prove that G-P-FRs do exist, which would confirm that the range of gravity is finite, and that the force of inertia is generated by interaction with distant masses which, in turn, would confirm that the speed of gravity is infinite. However, assuming that the nearest G-P-FR is located in the area of Local Void, that is, about 70 million light years from the earth, that experiment with spaceship won't be carried out. Thus, a direct confirmation of the EE theory's key implications appears to be difficult to obtain.

Finding direct confirmation of the existence of material energy *U* that fills the entire universe appears to be difficult as well. Yet, indirect confirmation of the existence of energy *U* could be easier to obtain as may be suggested from less-biased interpretations of LIGO experiment, chemical reactions, sun-planet interactions, etc. All proposed explanations of those physics mysteries involve material energy *U* and the equilibrium of *U* controlled by the execution of the SLT, which present the fundamentals of the EE theory. That is where the self-consistency of the EE theory can be claimed. Today's physicists believe in some energy, called "dark energy," to exist, filling the entire universe. That energy, however, isn't material as far as we know, as opposed to energy *U* which is ordinary matter. Whatever the energy that fills the universe is, it appears to be too feeble for the detection on its own. And yet, I am not sure of that. Haven't we detected it in hundreds of various experiments but refused to accept it for that wouldn't fit into the current, seemingly bias vision of the universe?

Furthermore, it is likely impossible to find direct evidence that nature doesn't do math, remember, measure quantities, etc. That is similar to the evidence problem that the CPP faces: It is likely impossible to find any direct evidence that nature realizes and measures space and time, solves equations, remembers things, etc.

53 Gravitational-potential-free space

The existence of G-P-FRs in the universe is possible only if the range of gravity is finite, which is a fundamental hypothesis on which the EE theory is based (section 1). The observed motions of cosmic and other objects present strong, while indirect evidence in support of a finite range of gravity. That is because without G-P-FRs, objects in an-infinite-in-space universe would be subject to forces of gravity generated by infinite amounts of mass in all directions. Yet, no interaction with an infinite amount of mass has ever been measured or observed.

54 Physical constants

According to FLH, laws of nature have to be nonquantitative for nature is dehumanized. That means that a law of nature cannot be executed according to a value of a constant built into the fabric of nature. That is for there can be no quantities built into the fabric of nature. *Physical constants are emergent properties of physical interactions*, the values of which emerge as a result of conflicts between the executions of the laws of nature (FLPs) and matter.

Because FLPs have no underlying principles, it isn't possible to derive a value of a physical constant from those laws only. An experiment/observation involving a measurement is necessary. In particular, the speed of light c being a constant value set by nature isn't possible according to FLH for nature doesn't realize space and time, whence the photon cannot own speed-measuring and speed-control devices. The well-established fact of experience is that about twenty physical constants currently are identified in physics, with not one of their values determined from theory. That is consistent with FLH and it appears to present indirect evidence in support of the EE theory. Constant D, which is the average mass (energy) density in G-P field, is possibly the only exception here. It appears to represent a true constant owned by nature.

55 Gravitational versus inertial mass

The standard belief is that two different masses exist, gravitational mass (m^g) and inertial mass (m^i), which are numerically equal according to GR's assumption. That belief was first put forward by Newton and accepted in CPP. According to the EE theory, on the other hand, m^g and m^i are one and the same thing (section 17). The problem with CPP's belief is that mass m^i is included in Newton's second law of motion (18), which is a physical law for *indirect* gravitational interactions, while m^g is included in law (9) applicable to *direc*t gravitational interactions. Because both those laws describe gravitational interactions, m^g and m^i are expected to be one and the same thing for there is only one fundamental law of gravitation (law II in section 3) that underlies the two physical laws, (9) and (18). Well-known experimental evidence for m^g and m^i being quantitatively equal strongly suggests that those masses are one and the same thing indeed.

The problem with the standard belief is that GR wasn't designed to address gravitational interactions with distant masses, whence mass m^i has no physical meaning in gravitational interactions addressed by GR (and Newton's law of gravity). It acquires such a meaning only if the inertia is assumed to be an innate property of mass, for instance, of a stone. That assumption, however, would humanize nature for the stone would have to make a decision on generating a force of inertia of an appropriate strength. On the other hand, no decision making is required in the EE theory. The strength of inertia force is set by the amount of conflict between a flying stone and distant masses.

56 Inertia

It seems that the prevailing concept of inertia among the physicists has been that the force of inertia (F_i) is the result of the interaction of a local mass object with distant masses (Mach, original thought of Einstein, Sciama, Brans and Dicke, and many others), which is a general statement of Mach's principle. The many mis-statements of that principle are excluded from this discussion. In general, it seems that Mach principle has acquired a bad reputation because Einstein attempted to incorporate it in GR, but it didn't work.

Mach's principle can be stated by the following relation,

$$F_i = f(m, distant\ masses). \quad (33)$$

where m is a mass of an object. According to the EE theory, the force of inertia is a result of the interaction of mass object m with distant masses as well (section 17), however, according to law II, it can have a non-zero strength in the presence of a conflict only,

$$\begin{gathered} F_i = f(m, distant\ masses), \\ F_i \neq 0 \text{ in the presence of conflict.} \end{gathered} \quad (34)$$

Consistent with law II, for an observable (quantitative) effect to occur, a conflict with the execution of an FLP has to exist. Let a car be idling at a red light. The fact of experience is that the force of inertia acting on the driver is nil, whence the prediction implied by Eq. (33) fails: While all matter pertinent to the generation of inertial effect is there—that includes mass m (the car) and distant masses—the force of inertia is nil. That suggests the premise of the execution of a law of gravitation in itself causing a quantitative effect to be incorrect. Relation (34), on the other hand, correctly predicts the force of inertia to be nil as no conflicts affecting the interaction of mass m with distant masses exist. Let the light turn green. The car accelerates. A conflict is inflicted and the force of inertia acting on the driver becomes non-zero in agreement with relation (34). A conflict means that the degree of equilibrium between the car and distant masses changes.

The same argument for the viability of the EE theory can be stated in more fundamental terms. According to CPP's view, the fundamental law of gravitation is: A mass object quantitatively interacts with all other mass objects in the universe. According to the EE theory, that law is stated as follows: A mass objects has the potential to interact quantitatively with other mass objects located within its G-P-FH. The evidence supporting the suggested origin of the force of inertia—which is based on Mach's idea supported by FLH and a finite range of gravity—appears to be vast: Every stone laying on the ground, every bullet in a rifle, every apple hanging from a tree, and every driver in an idling car that we observe *has a potential only to interact* quantitatively with distant masses. Inflicting a conflict between objects (forcing a change in the degree of equilibrium) is necessary for a quantitative effect to occur, that is, for the real force of inertia to emerge. That appears to present strong evidence in support of FLH.

In agreement with CPP, distant masses cannot be the source of inertia as the speed of gravity is assumed to be finite, while the well-confirmed fact of experience is that the force of inertia is generated simultaneously with inflicted conflict. In the EE theory, the speed of gravitational interaction is infinite (section 11), and the immediate effect of inertia—e.g., feeling the force of inertia in accelerating car—presents strong evidence for that speed to be infinite.

57 The speed limits for material and immaterial entities

The speed of gravitational interaction was suggested to be infinite (section 11). It is also suggested that the speed of a material object, which comprises material energy *U*, is finite. That is for the speed of energy *U* (gravitational wave), released as a result of the motion of an object toward equilibrium (law Ia), is expected to be finite and equal to the speed of light *c*, as discussed in section 44. (That is the reason that an apple doesn't fall from a tree instantly.)

It is often said that the speed of gravitational interaction must be finite for that is a requirement of the special theory of relativity (SR). This is a notion that appears to be unsupported. In physical terms, SR doesn't imply any speed limits on either material or immaterial entities. The limit on the speed of a material object that comprises energy *U* is consistent with SR when accounting for the when accounting for the mathematical concepts of the Lorentz factor and the relativistic mass. That limit implies that no material ("mass-energy") objects can travel faster than the speed of light *c*. In terms of material objects, therefore, the proposed EE theory is supported by the mathematics of SR.

The celebrated limit on *the speed of immaterial interaction* (the speed of immaterial information) results from the consideration of Minkowski's spacetime, which is referred to as the causality proof. From the perspective of FLH, however, Minkowski's spacetime is an abstract construct that doesn't exist in physical terms, which means that it doesn't exist outside the physicist's mind. As a result, the causality proof isn't really a proof. Furthermore, dehumanized nature is unable to realize space and time, which implies that the speed of interaction is expected to be instantaneous. Dehumanized nature can own no laws that have spatial or temporal restrictions.

58 The direction of the force of gravity

As suggested in section 36, the direction of the force of gravity acting on a planet orbiting the sun is the direction of the orbit. That implies that the direction of the force of gravity is an emergent property of gravitational interaction rather than a pre-set direction of the line connecting gravity centres of interacting objects. The latter view humanizes nature, for dehumanized nature is unable to determine the locations of objects' gravity centers as she owns no rulers nor weight scales. The penalty the latter view pays is that some observable, gravitational interactions have to be described in terms of fictitious forces that don't exist in a physical sense.

The physics underlying the suggested, emergent direction of the force of gravity is that it must act in the direction of motion. That is the direction of the force of inertia that acts in the opposite direction, such that the action-reaction law is obeyed. Every car driver accelerating toward north or south, or in any other direction, proves that the force of inertia acts along the direction opposite to inflicted conflict, that is, in the direction opposite to the force that imposes motion. Therefore, if it is a force of gravity that imposes motion, it has to act in the direction opposite to the force of inertia, and not in the direction pre-set by the gravity centers of interacting objects. This suggestion concerning the direction of the force of gravity appears to provide evidence for the viability of the EE theory as the fictitious forces doing real work can be eliminated. In particular, the existence of a centripetal force as a physically real force can be eliminated (section 36), with the direction of the planet-sun force of gravity emerging as a result of the planet-sun equilibrium condition.

59 The classical and the quantum physics levels

The electron-nucleus, gravitational-time dilation, stellar coronae, sun-planet interaction, pendulum clock and some other scenarios discussed throughout this book suggest that nature is all about material energy U and energy equilibrium, which are the basics of the proposed EE theory. Those scenarios suggest that, it is the SLT that single-handedly runs the universe while the action-reaction law always is obeyed such that the universe is in equilibrium in each instant of time. Those scenarios also suggest that the same FLPs exist, and are executed in the same way at both the classical and the quantum physics levels. That suggestion is consistent with the implication of FLH: As said before, because nature is unable to realize space and time, she cannot know if it is nanoseconds and angstroms or hours and meters out there at the place of a physical phenomenon she is attending. The physics at the subatomic level cannot be different from physics at the cosmic level for nature has been dehumanized.

60 Quantum theory

Quantum theory (QT) is highly successful. Physical laws for direct and indirect gravitational interactions expressed by Eqs. (9) and (18)—both due to Newton with some new interpretations and modifications—are highly successful as well. That supports the EE theory if both theories are examined from the perspective of the FLH corollary. While that corollary is a part of the EE theory, it also is meant to apply to the formulation of any other physical law. The two proposed laws for gravitational interactions, (9) and (18), follow the FLH corollary in the following ways:

- Laws (9) and (18): They prevent nature from recognizing a force field that has a property variable with location or time, which would be the kind of force field normally used in applied physics. Contrary to the physicist, nature is unable to measure or calculate the amounts of physical quantities. As G-P field introduced in section 8 has no variable properties, an allowance for that inability of nature has been made in the two laws for gravitational interactions. G-P field included in laws (9) and (18) has no quantitative properties.
- (Law 9): It prevents nature from knowing that one of two interacting objects contributes to the force of gravity more than the other object. That was called inference (a) in section 9. Nature is unable to measure and compare the amounts of

contribution. That also implies that the structure of law (9) prevents nature from knowing the amount of mass of an interacting object, which is a consequence of FLH (nature has no weight scales).

- Law (18): It prevents nature from knowing the amounts of distant masses. Nature is unable to measure and thus know an amount of mass. If the supercluster illustrated in Fig. 5 were two times more massive, the acceleration generated in response to the application of force F_m of a certain strength would still be the same. The average mass density D in G-P field is what nature responds to. That is consistent with the suggestion made in section 54 concerning the average mass density D being a constant of nature.
- Laws (9) and (18): They prevent nature from realizing the direction of the force of gravity. As nature doesn't realize space, she is unable to set up a direction in which an emergent force would be acting. The direction of a force is set by the direction in which the action-reaction law has to be obeyed, that is, the direction along which the inertia and the gravity forces balance out. That always is the direction of motion.
- Law (18): It prevents nature from realizing and accounting for the distance to distant masses. Nature is unable to realize distance (space). The amount of acceleration depends on the amount of conflict only, which is the strength of the applied force F_m. It doesn't depend on the distance to distant masses.

Consistent with the FLH corollary, the primary notions of QT dehumanize nature by making allowances for her inabilities in the following ways:

- Sum over histories: A particle emitted at point A can travel over any of an infinite number of pathways before it arrives at point B. That is clearly impossible. An electron emitted at point A toward point B will travel in a straight line between the two points if undisturbed. Yet, the sum over histories supposition contributes greatly to the success of QT. It seems that in that supposition, the FLH corollary was accounted for: Nature is unable to predict or know the locations of a particle in space and time (the travel path) as she doesn't realize space and time. A physical law should make an allowance for that inability. And that is what the sum over histories hypothesis does: As nature cannot possibly know the path of a particle, she has to be forbidden to know it. That is forbidden by the possibility of a particle following any path.

- Wave function: In a closely-related supposition, a particle in an instant of time can be located anywhere in space, and only the probability of a particle being at a specific point in space can be known. Thus, there are non-zero probabilities that a photon emitted from Paris toward Warsaw will be detected in Madrid. That, again, isn't possible. The concept of a wave function appears to follow the FLH corollary by forbidding nature to know the position of the photon before the position is measured. Again, the concept of a wave function is central to the success of QT.
- Quantum fluctuations: As energy U fills the entire universe, the material-energy-conservation law wouldn't be violated by the creation of virtual particles.[r] However, spontaneous work by nature mustn't lead to an increase in nonequilibrium of a system (law Ia), which would happen if particles were to be created. Therefore, quantum fluctuations aren't possible. Yet, nature is unable to know if a particle was or wasn't created because she can have no memory of a prior instant of time in which the particle would, or wouldn't exist. Forbidding the creation of particles would imply that nature has a memory as she could tell that no particle was there in a prior instant.

The above QT suppositions are often said to be weird. It also may seem weird that nature cannot know the difference between the mass of an apple and the mass of the earth, as is suggested by physical law (9). From the perspective of FLH, however, there is no weirdness here. Those suppositions reflect rational reasoning: In formulating a physical law intended to describe nature, one should forbid nature to do something that she cannot possibly do.

---

[r] Some physicists think that quantum fluctuations (creation and annihilation of virtual particles) violate energy-conservation law, which is permitted because the fluctuations are short-lived such that they cannot be observed. As the entire universe is filled with energy U (section 21), the material-energy-conservation law cannot be violated according to the EE theory.

In other words, one should dehumanize nature. The physical laws for gravitational interaction and QT—appear to be based, at least in part, on the same physical law development rule, stated in as the FLH corollary in section 2.

The standard view is that QT doesn't account for gravity. It is difficult to accept that view for gravity appears to be a keystone of physical interactions. Said another way, it appears impossible that a theory as comprehensive and successful as QT could be formulated without accounting for gravitation. In that regard, it needs to be appreciated that QT is merely a mathematical model of subatomic interactions and, as such, doesn't have to account for gravity explicitly. It is sufficient to demonstrate that QT isn't in disagreement with observable gravitational effects. Regardless, it has been suggested that G-P field plays the role of a mediating agent in both electric and gravitational interactions (section 30), which provides a link between QT and gravity. In section 33, it was suggested that gravity is, in fact, accounted for in QT for the stability of a nucleus is maintained by the gravity forces that act between nucleons. If spin and charge are the same physical effect, as tentatively suggested in section 32, that would provide further evidence that QT incorporates gravity.

61 General relativity

GR is a highly successful physical law designed to describe direct gravitational interactions. It isn't intended to describe interactions with distant masses. The proposed EE theory (abbreviated "EE" for the purpose of this section) is intended to describe both direct gravitational interactions and interactions with distant masses, including the force of inertia. EE and GR are principally different for in EE, immaterial (abstract) quantitative spacetime is replaced with a physically-real material energy U and G-P field. Moreover, EE claims the range of gravity to be finite and the speed of gravity to be infinite, the principal ideas that are contrary to GR's ideas. A comparison of the main features of GR and EE is important to this book. It is summarized as follows:

- The SLT: In EE, the fundamental law of gravitation (law II) presents a straightforward aspect of the SLT. It follows that gravitational interactions obey the SLT by default. In GR, the

SLT doesn't relate to gravitational effects (it doesn't affect the equivalence principle), whence the SLT isn't disobeyed.

- The action-reaction law: In EE, gravitational interactions obey the action-reaction law since the force of gravity, or a component of it, must be balanced by the real force of inertia. In GR, a force of gravity doesn't exist (there are no forces in GR), whence GR doesn't disobey the action-reaction law.
- Principle of equivalence: Both GR and EE account for the equivalence principle. In GR, the equivalence principle is an exact law of nature that has no underlying principles. In EE, the equivalence principle is a physical law that results from FLH, and it isn't exact owing to a finite range of gravity.
- Interaction with distant masses: GR doesn't account for the force of inertia that results from interactions with distant masses. (Gravitational interactions with distant masses are imperceptible in GR.) In EE, the force of inertia, which is generated by the interactions of objects with distant masses, is a key feature of the theory. It cannot be imperceptible for it has to balance any motion-imposing force such that the action-reaction law is obeyed.
- Material energy: In EE, material energy U (matter) plays a key role as it fills the entire universe and is inherently related to equilibrium. In GR, energy as a material substance doesn't exist. The universe between objects, including atomic space between nucleons and electrons, are empty of matter.
- Material energy U and spacetime: In EE and GR, energy U and spacetime, respectively, fill the entire universe. In EE, energy U spreads out in the form of gravitational waves that are material. In GR, the disturbance of spacetime spreads out in the form immaterial (abstract) gravitational waves.
- Gravitational and inertial masses: In GR, gravitational and inertial masses are numerically equal. In EE, gravitational and inertial masses are one and the same thing.

CHAPTER X – THE UNIVERSE

The conceptual cosmological model of the universe, the highlights of which are discussed in this chapter, is based on the assumptions and the conclusions of the EE theory. It is a model, which will be called EE cosmological model, in which the universe is infinite in space and time.

62 Cosmology

In today's cosmology, astronomic observations and measurements are normally interpreted in accordance with the assumptions of the cosmological (big-bang) model which, at this time, is considered "standard." Other interpretations are rarely pursued. The central feature of the big-bang model is the universe that was born at some time in the past having infinitesimal size, and has been expanding ever since.

Because nature doesn't own clocks and rules according to FLH, she is unable to realize space and time. Thus, it was concluded in section 4 that nature cannot embrace any space or time restrictions on her laws. Hence, the universe must be considered to be infinite in human time and human space. In that regard, one needs to keep in mind that according to the EE theory, nature comprises matter (energy U) and the laws of nature with no indication that either of those was, or could have been created at a point in time.

It appears that the standard cosmological (big-bang) model has been widely accepted because of two observational-astrophysics giveaways. Shortly after 1929, the big-bang proponents claimed cosmological redshifts to be an undisputable evidence supporting the cold big-bang model. Shortly after 1965, they claimed cosmic microwave background (CMB) to be an undisputable evidence supporting that model. In this chapter, other interpretations of cosmological redshifts and CMB are put forward, primarily from the consideration of FLH.

63 Infinite universe

In explaining specific physical phenomena in the prior chapters, it wasn't essential to know if the universe was finite or infinite in time. To explain how the universe works under the dehumanized nature scenario, however, a finite- or infinite-in-time universe has to be either concluded or assumed. We either live in a special point in human time or not. As suggested in section 4, the universe should be considered infinite in human time and human space as nature has been dehumanized. That was a suggestion derived from FLH, and it is consented in the following discussion.

The sun is shining. Neither the sun nor any other star can stop "shining," that is, stop emitting material energy U. Otherwise, given infinite human time, there would be an infinite number of stars that don't emit energy (matter), which is in disagreement with the observations. As a star never stops emitting energy U, each star must eventually die given infinite time, which means that it must disappear by emitting its entire energy U. It follows that, on the average, for any star that dies, another star has to be born. Otherwise, given infinite time, all stars would have already died, which isn't the case. On the average, a star must emit more energy U than it absorbs so that there is matter available for the formation of new stars (the material-energy conservation law holds).

A star that is dying by emitting material energy U, which means that the matter it comprises is being dispersed, causes an increase in the entropy of the universe because the disorder in the universe increases. A star that is born out of energy U causes a decrease in the entropy of the universe because the disorder decreases. The total entropy in a representative volume of the universe can neither increase nor decrease for the universe to be infinite in time. It must remain constant, on the largest scale, which is consistent with the suggestion that for each star that dies, another star has to be born. Otherwise, given the infinite-in-time universe, the entropy in the universe would be infinite or zero, neither of which is the case.

Why is it that according to the big-bang cosmological model the entropy in the universe increases? The answer appears to be: Some of the radiation emitted by a star is never absorbed by other stars or used in the formation of new stars. Thus, there is more energy (matter) dispersion than accumulation in the universe, which leads to a continuous increase in the entropy of the universe.[s]

---

[s] That some matter (radiation) emitted by a star would never be absorbed by other stars was one of Einstein's concerns with Newtonian gravitation. He believed that "... the [Newtonian] universe should have a kind of centre in which the density of the stars is a maximum, and that as we proceed outwards from this

A universe that is infinite in time must also be spatially infinite.[t] That is because given infinite time, all objects in a spatially-finite universe would have already collapsed to its gravity center, which disagrees with observations. (This is the argument originally put forward by Newton in support of a spatially-infinite universe. Some modern cosmologists reject that argument.) Note that in the spatially-finite universe, the forces of gravity acting on an object cannot be infinite even if the range of gravity is infinite, such that cosmic and other objects can be in motion. All objects actually are expected to be in motion as discussed at the end of section 18. Another argument against the expansion of a spatially-infinite universe is that it has nothing to expand into.[u]

If the universe is finite in time and came into existence at time t_0, then the laws of nature and material energy U, which constitute nature, would have to come into existence at time t_0. Therefore, time t_0 would have to be known to nature before time t_0. (To weasel out of this problem, an assumption typically is made that time hadn't existed before time t_0.) That would contravene FLH for nature before time t_0 would have to exist without nature's laws and material energy U, that is, *without herself.* That means that a creator (God) would have to exist to create nature, space and time according to God's intent and designs. That means that God would

---

centre the group-density of the stars should diminish, until finally, at great distances, it is succeeded by an infinite region of emptiness." (Relativity: The Special and General Theory, 97, Penguin Books, 2006.) That premise led Einstein to expressing a closely related concern, "... [the Newtonian universe] is still less satisfactory because it leads to the result that the light emitted by the stars [is] perpetually passing out into infinite space, never to return, and without ever again coming into interaction with other objects of nature." (Ibid, 98). Both those concerns disappear as soon as it is agreed that the universe is infinite in time and in space, while the range of gravity is finite.

[t] The idea of spatially-infinite universe is very old. According to Thomas Kuhn: "In the fifth century B.C., the Greek philosophers, Leucippus and Democritus, visualized the universe as an infinite empty space, populated by an infinite number of minute indivisible particles or atoms moving in all directions. The earth was one of infinite, essentially similar heavenly bodies formed by the chance aggregate of atoms." The idea of spatially-infinite universe had prevailed in western science until the big-bang model was proposed in the early 1930s.

[u] In the standard cosmological model, the "nothing-to-expand-into" inference is avoided based on the assertion that curved spacetime exists as a physical reality, which can make the universe finite in size, with no boundaries. In the universe with no boundaries, there is no space outside the universe, whence its expansion does not require something to expand into.

have to be humanized. In the scenario in which the universe is finite in human time but God doesn't exist, non-existing nature would have to decide, prior-to-creation, on the particulars of the act of creation, that is, on the amount and the kind of matter that she would create and what laws she would create, while she is unable to make plans and decisions according to FLH (section 2). While space and time are physically real from the perspective of the human mind (section 4), nature doesn't know that those exist such that she, by herself only, couldn't possibly create a finite-in-time-and-space universe. Humanized creator (God), it appears, is the only rescue that can support a finite universe if FLH is to hold true.

To recap the above arguments for an infinite universe: As nature doesn't realize space and time, *no laws of nature executed with the consideration of space and/or time can exist.* On the other hand, such laws would have to exist if the universe were to be finite in human space and/or in human time. Said another way, unless the universe is infinite in human space and human time, nature would have to know when and where she—that is, her laws and matter—can or cannot exist. That knowledge is forbidden by FLH as nature owns no rules and clocks, nor memory sticks. The bottom line is: If the FLH holds true, the universe has to be considered infinite in human space and in human time.

The arguments in favor of infinite universe may appear weak for they are purely philosophical. Yet, from the perspective of FLH those arguments are very strong for in a finite universe nature has to be humanized, which means that FLH couldn't hold true.

I want to emphasize again that if FLH holds, the arguments for a finite universe have to include the existence of a creator of the universe (God). Nature cannot be the creator by herself for she mustn't be humanized according to FLH. An act of creation of the universe has to involve decision making, usage of memory, design board, plus the unconceivable ability to make something out of nothing. That means that the creator (God) has to exist and be humanized. Under that circumstance, the universe can be finite.

The humanization of Gods goes back not only to the ancient Greeks as discussed in Preface. It was later followed in Hebrew bible, the authors of which wrote: "And God said, let us make man in our image..." (Genesis, 1:26). It appears that the authors of the bible got it upside down. It actually was them who made God in man's image, and not the other way around.

Finally, it needs to be said that Leibniz's principle of the identity of indiscernibles applies well to the suggested idea of an infinite-in-space universe. That is for no two objects can be subject to exactly the same radiation beams (comprising matter) coming-in from all directions of the universe. That will be further stressed in section 65 after the locally nonhomogeneous distribution of G-P-FRs in the universe is conjectured, and in section 68 where the influence of those inhomogeneities on the spectra of cosmological redshifts—which means, on the energy U_{photon} carried by the light beams—is discussed.

64 Cosmic voids

Cosmic voids are the regions of the universe that contain very few or no galaxies, as opposite to the galaxy filament that surrounds each cosmic void. The cosmic voids-galaxy filament structure of the universe was discovered from the studies of redshift surveys in the late 1970s through early 1980s by Joeveer et al.,[79] Gregory and Thompson,[80] Kirshner et al.,[81] Huchra et al.,[82] and others, as recapped by Bond et al.[83] It was an outstanding discovery in astronomy, and the investigations of void galaxies have become a key astrophysics research subjects ever since. A summary of that research is provided by Weygaert and Platen.[84]

As the densities of mass-energy in cosmic voids are observed to be very low, it appears rational to identify cosmic voids with gravitational-potential-free regions (G-P-FRs) of the universe. As pointed out in section 12, G-P-FRs have to exist in the universe. In the big-bang model, "empty space" inside a cosmic void is no different from "empty space" in galaxy filament. In the EE model, those two "empty spaces" are substantially different: *Space in a cosmic void contains no consequential G-P field.* And, no direct force of gravity penetrates the boundary of a cosmic void.

A support for the identification of G-P-FRs with cosmic voids comes from the proposed interpretation of temperature (section 28) and the actual CMB-based measurements of temperature in superclusters and supervoids (see, e.g., Granett et al.[85]). Those measurements showed supervoids to be distinctly "colder" than superclusters. As cosmic voids, herein identified with G-P-FRs, are expected to contain relatively low densities of energy U_{fluid}, their average temperatures are expected to be lower than those in the galaxy filament comprising superclusters, which agrees with the results of CMB-based surveys.

65 Large-scale gravitational structure of the universe

The large-scale homogeneity and isotropy of the universe and the existence of regions that contain no consequential G-P field imply two possible structures of the universe. In the first scenario, the universe comprises a web-like G-P field, which will be called galaxy filament, extending to infinity and filled with bubble-like G-P-FRs (cosmic voids—see section 64). On a large-scale, cosmic voids are distributed homogeneously throughout the universe.

In the second scenario, it is the opposite. The structure of the universe comprises a web-like gravitational-potential-free space, which extends to infinity and contains bubble-like G-P fields distributed homogeneously throughout the universe. The problem with the latter scenario is that given infinite time, the masses contained in each G-P field bubble would have long collapsed to its gravity center, which hasn't happened. Moreover, under that scenario the large-scale-gravitational composition of the universe would be discontinuous, whence it would be difficult to argue that the same laws of gravitation apply to each bubble, that is, to the entire universe. Hence, it is suggested that it is the former scenario that appropriately describes the large-scale-gravitational structure of the universe, as illustrated in Fig. 26. Those arguments in favor of the former scenario aren't rigorous in any way. However, that scenario appears to fit well the observed features of the universe comprising cosmic voids and the galaxy filament.

In the suggested structure of the universe, the galaxy filament and the cosmic voids are separated by the boundaries of the G-P field of the filament. However, G-P fields of stars and galaxies are also expected to exist inside cosmic voids (that expectation will be elaborated on in section 69). Those fields would be local, that is, entirely separated from the G-P field of the galaxy filament by gravitational-potential-free space. Typical diameter of a cosmic void is about 1×10^8 ly (e.g., Foster and Nelson[86], Pan et al.[87]).

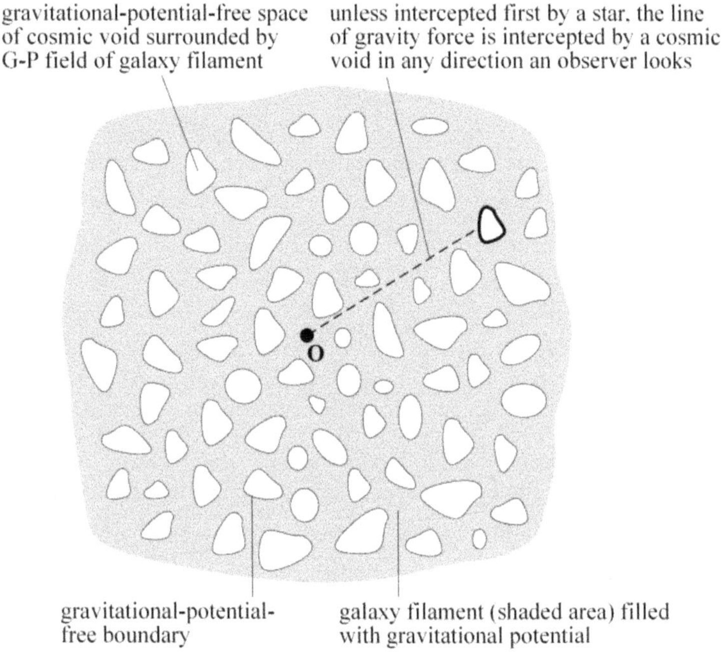

Fig. 26 Highly schematic large-scale-gravitational-structure of the universe. It comprises continuous G-P field (galaxy filament) filled with numerous galaxies/stars. G-P free cosmic voids contain none or very few galaxies.

Using postulate (1), the "diameter" of the G-P field of a typical galaxy, having a mass of two hundred billion solar masses, is estimated to be between 1.4×10^4 ly and 1.3×10^7 ly, depending on the actual average mass density D in G-P field. (The range of likely D values will be discussed in section 67.) The above range of cosmic void galaxy diameters has been estimated assuming spherical rather than "disk-like" galaxy shapes, which presents a rough approximation only. Nonetheless, it is sufficiently accurate to conclude that there would be ample space in a typical cosmic void, as delineated using cosmic survey results, to contain G-P fields of many stars and galaxies, if the actual value of density D is closer to the upper bound estimate of 3.0×10^{-19} kg·m^3 than the lower bound estimate of 4.0×10^{-28} kg·m^3.

At first glance, one might expect the G-P field – no-G-P field boundary that delineates a cosmic void to be an observational feature of the universe. On a closer examination, however, one has to conclude that there is nothing conspicuous about the boundary of a cosmic void. It is merely a boundary of the posted immaterial information on the existence of gravitational potential (section 8). Consider a star situated inside a cosmic void and moving toward the boundary of the void, eventually entering the G-P field of the galaxy filament. As the star passes the boundary, the indirect force of gravity and the force of inertia acting upon it appear, first at extremely small strengths. The star "feels" no substantial change in the strength of those forces, and nothing spectacular happens. A boundary of G-P field doesn't comprise any matter. As such, it doesn't interact with anything and it doesn’t generate any effects that could be observed.

This book started with the suggestion that the assumption of an infinite range of gravity, put forward by Newton, was problematic for it resulted in infinite amounts of mass acting on objects from all directions (section 1). The large-scale structure of the universe schematically illustrated in Fig. 26 removes that problem: There is no place in the universe where an object could be subject to the interaction with infinite amount of mass as any line of force, for instance, the dashed line illustrated in Fig. 26, has to have a limited length for it has to eventually intersect the boundary of a cosmic void. (That corresponds to the shooting-rifle-in-forest scenario discussed in footnote e.) The star illustrated in Fig. 5 is completely surrounded by its G-P-FH, which means that there is a limited amount of mass in any direction that the star can interact with.

The large-scale structure of the universe starts with the U_{solid}-core particles: the electron, the proton, and the neutron, and their combinations that form matter arrangements such as planets, stars, atoms, etc. *Each of the U_{solid}-core particles and each arrangement comprising those particles, must have a consequential and well-defined G-P field*, which is a prerequisite for the existence of a g-object. In relation to the suggested structure of the universe, only well-defined G-P fields may have G-P field boundaries.

66 Dark matter

In the universe that is infinite in time, the very old and very cold (VO&VC) stars and galaxies are expected to be abundant. On a large scale, those have to be homogeneously distributed over the universe in accordance with the cosmological principle. The abundance of VO&VC stars and galaxies is suggested because the rate of star's cooling down—which is expected to be proportional to the rate of energy *U* emission—decreases with the temperature of the emitting body to the power four. That is a conjecture derived from the Stefan-Boltzmann law. If this conjecture is about correct, the number of VO&VC galaxies in a representative volume of the universe would be orders of magnitude greater than the number of galaxies that emit radiation in the visible light frequency range. If the VO&VC stars and galaxies exist indeed, it is rational to posit that those constitute the dark matter discovered by Zwicky[88].

Dark matter comprising VO&VC stars/galaxies implies that it is the ordinary baryonic matter comprising atoms, that is, the electrons, the protons, and the neutrons, and material energy U. In current physics, dark matter is believed to be a non-baryonic matter that doesn't emit radiation. The emission of the photon discussed in section 33 appears to be conceptually simple in the case of ordinary atomic matter. Matter that doesn't emit radiation, on the other hand, has never been discovered and there seems to be no indication that such matter could comprise atoms as we know them. The physics of non-baryonic dark matter is unknown and without precedence, while the prospect of direct detection and testing of such matter appears to be bleak.

Astronomic observations suggest that the majority of visible galaxies are grouped in gravitationally bound clusters. If it is gravity that forms galaxy clusters, then it is reasonable to expect that a typical cluster should have a roundish shape just like a typical galaxy does—that is, the structure of a cluster should be similar to the structure of a galaxy. The roundish shapes of many galaxy clusters have been observed (Pascal de Theije et al.[89]). Furthermore, it can reasonably be expected that the number density of galaxies should increase toward the center of a galaxy cluster just as the number density of stars increases toward the center of a galaxy. That is how gravity is expected to work over long time: It pulls-in all stars that form a galaxy together, and pulls-in all galaxies that form a galaxy cluster together.[v]

---

[v] The clustering of galaxies is often described in terms hierarchical clustering, which means that larger structures in the universe are formed by the mergers of smaller, structurally similar structures. That conceptual description can also

If the density of visible and invisible galaxies in galaxy clusters is distributed in a pattern similar to the density of stars in visible galaxies, then it can be expected that the concentrations of both visible and invisible galaxies are highest at the centers of galaxy clusters. Said another way, the observed clustering of visible galaxies should correlate with the clustering of VO&VC galaxies.

67 Average mass-energy density

There doesn't seem to be any hard, unbiased data that can be used to estimate the average mass density in the entire universe, or in G-P field. The expected abundance of invisible VO&VC stars and galaxies implies that the actual mass density in the universe is orders of magnitude higher than the density that can be estimated from the observations of visible matter only.

In the case of a single galaxy, the invisible matter is believed to comprise dark-matter galactic halos (e.g., J. Bahcall et al.,[90] N. Bahcall and Kulier[91]). From the studies of galaxy rotation curves, the total amount of matter in a galaxy is estimated approximately ten times larger than the amount of visible matter (e.g., Bennet et al.[92]). From the perspective of this discussion, the widely-quoted "ten times larger" factor could be a low estimate for it is based on Newton's law of gravitation (2), which means that function *K* in law of gravitation (9) would remain of equal value at any distance from the center of a galaxy. Another problem with the "ten times larger" factor is that it has been found to apply to the very young (visible) galaxies. Without the benefit of actual observations and measurements, that shouldn't be automatically extrapolated to all galaxies, particularly to the VO&VC (invisible) galaxies that are expected to prevail in the universe.[w]

The currently estimated average mass density in the universe is 1.0×10^{-26} kg·m^{-3}. It is a bias estimate because it corresponds to the critical mass density, which is a mathematical concept applicable specifically to the big-bang model. The critical mass density was estimated from the interpretations of sky surveys (e.g., WMAP

---

apply to a universe in which invisible galaxies are abundant, as in the proposed EE cosmological model.

[w] It seems somewhat immature to think that the visible galaxies must prevail in the universe just because we can see them. Perhaps it is just the other way around and we see them because they affect the life of a human (section 5).

and BOSS surveys). The 1.0×10^{-26} kg·m^{-3} estimate is normally believed to account for dark matter and hypothetical dark energy.

The density of visible matter in the universe is estimated at about $D_{\text{lower}} = 4.0\times10^{-28}$ kg·m^{-3}. That value presents an unbiased, *lower bound* mass density estimate as it was derived from actual observations, regardless of any implications of a cosmological model or theory. The expected existence of dark matter in the form VO&VC stars and galaxies suggests that this estimate of the total, actual mass density in the universe is unrealistically low.

The *upper bound* value of the average mass density in the G-P field (D) can be estimated from the fact that the volume of the G-P field of Milky Way galaxy cannot be less than the volume of its visible disk. From the approximate amount of mass contained in Milky Way (2.0×10^{42} kg)[93] and a rough volume of the galaxy visible disk (6.6×10^{60} m^3), the upper bound value of the average mass density in G-P field can be estimated at $D_{\text{upper}} = 3.0\times10^{-19}$ kg·m^{-3}. That estimate ignores possible existence of dark matter (VO&VC stars) inside the Milky Way disk.

68 Cosmological redshifts

It follows from the prior discussions that what we really know about cosmological redshifts of starlight-beams measured in earth observatories is this: A photon comprising a quantum of material energy U_{photon} is emitted from a far-away galaxy and, after a long journey through space, it arrives at a laboratory comprising less energy, which means that the photon has been redshifted because $U_{\text{photon}}^{\text{observed}} < U_{\text{photon}}^{\text{emitted}}$. Thus, the photon has lost some of its energy U_{photon} because of its gravitational interactions. According to law Ia, the photon loses some of its energy because it is driven by the execution of the SLT toward higher of equilibrium (section 44). The conclusion therefore is: *At least a portion of the measured cosmological redshift must be due to the loss of material energy* U_{photon} *(dimension [M]) the photon comprises*. It follows that the photon must also lose some of its ability to do work

In 1929, Edwin Hubble published his famous paper,[94] in which he derived a radial-velocity – cosmological-redshift relationship for visible galaxies. Later, he called that relationship "the law of red-shifts."[95] The standard interpretation of Hubble's findings is that he discovered galaxies to recede from each other with the recession speeds proportional to the distances between them. That interpretation became the pillar of the act-of-creation-expanding-universe-cosmological (the big-bang) model. The big-bang model is typically accepted as a "carved-in-stone" cosmological model as evidenced by the statements of many prominent physicists (e.g., R. Penrose,[96] F. Wilczek[97]).

Yet, the standard interpretation of redshifts on which the big-bang model is based (redshifts are entirely due to Doppler's effect) isn't substantiated. Hubble couldn't possibly analyze the speeds of galaxies because he didn't measure any galaxy speeds. What he measured were the redshifts of starlight beams. In referring to the radial velocities of galaxies, Hubble *assumed*, likely influenced by the proposal of V. Slipher[98], that redshifts were due entirely to Doppler's effect. (He later had serious doubts concerning that assumption[99].) It is important to realize that Hubble wouldn't be able to measure galaxy velocities even if he intended to. That is plainly explained by the statement of Smithsonian Astrophysical Observatory: "Galaxies are ... so far away, that you could never see them move just by looking—even if you looked for a whole lifetime through the most powerful telescope!"[100] That statement is applicable to linear motions of galaxies in general, and to radial motions in particular.

A few months after Hubble's paper was published, Fritz Zwicky proposed an alternative explanation of cosmological redshifts[101] that was later named the "tired-light hypothesis." (In recent years, several specific tired light explanations of cosmological redshifts were proposed by a number of renowned scientists, including P. A. Violette,[102] J. C. Pecker et al.,[103] J. C. Pecker and J. P. Vigier,[104] M. Harney,[105] J. F. G. Julia,[106] A. Ghosh,[107] and others. A comprehensive listing of the various proposals put forward to explain the cosmological redshift effect has been presented by Louis Marmet.[108]) The key idea behind the tired-light hypothesis is that the photons lose their energy (dimension $[M \cdot L^2 \cdot T^{-2}]$) owing to the interactions with other particles or media while traveling over the vast distances of the universe.

In the EE cosmological model, the tired-light concept is introduced based on the consideration of a finite range of gravity (sections 1 and 8), the photon being a quantum of material energy

U (section 24), law Ia (section 3), and the energy equilibrium of the photon – G-P field system (section 44).

The photon in motion through the G-P field is driven by the execution of the SLT. In accordance with law Ia, material energy *U* (dimension [M]) is released for the photon is driven toward equilibrium with energy *U* contained in the G-P field (section 44). As gradient *D* has a constant magnitude and the photon moves under a steady-state process, the amount of released energy *U* is expected to be proportional to photon's travelled distance (section 44), which is a supposition consistent with the general idea of the tired-light hypothesis. In this proposal, however, it is suggested that the energy released by the photon is material.

One shouldn't assume that cosmological redshifts are due to the losses of photon's energy only, and unaffected by the radial speeds of starlight-emitting galaxies. That is for the Doppler-shift effect applies as much to waves as it may apply to any other periodic effects, for instance, the emission of the photons.

The interpretation of cosmological redshifts proposed here is that the effect of the release of material energy *U* by the photon overwhelms the Doppler's shift effect for far-away galaxies. That means that galaxies could be moving toward us or away from us (thus have some "peculiar" velocities), but we would always observe redshifts for the galaxies that are so far away that the loss of energy *U* effect overwhelms Doppler's shift effect. That is consistent with Hubble's original findings. He found that, out of the twenty-four galaxies he studied, the spectra of the four closest to the earth galaxies were blueshifted, except for Magellanic Clouds, which indicated that those four were moving toward us. (Hubble also found a fifth blueshifted galaxy located farther away, but its blueshift was very small.) That is consistent with the proposal that the loss of energy *U* effect overwhelms Doppler's shift effect at larger distances, at which cumulative loss of material energy is larger. Cosmological principle implies that, in a radial direction, about half of all galaxies are moving away from us, and the other half are moving toward us. Nonetheless, if the proposed explanation of cosmological redshifts is correct, one would expect that the larger the distance to the source of a starlight beam is, the less of a chance there is to observe a blueshifted starlight beam, which is what Hubble found.

A number of famed astrophysicists suggested in the past that the Doppler-shift effect was responsible partially only for the actual, observed cosmological redshifts. Usually, that suggestion was made with reference to the redshifts of quasars (e.g., H. Arp[109], J. V. Narlikar[110], H. Arp et al.[111]). It is worth noting that J.-C. Pecker, another famed astronomer, did identify problems with the Doppler-effect-only interpretation of cosmological redshifts, with reference to the observed spectra of light beams emitted by the sun, binary stars and close-by galaxies.[112]

Most of the galaxies that are observed using optical telescopes emit similar-energy photons. Hence, about the same cosmological redshift–distance relationship would be expected to hold. And it does. However, the same relationship might not be well suited for all cosmic objects. With reference to the prior paragraph, it could be different for quasars. A photon emitted from an extremely massive object, such as a quasar, or a star at the center of a galaxy cluster, would initially be subject to a high *gravitational redshift* and lose a relatively large amount of energy shortly after the emission. That means that some quasars could actually be located closer to the earth than currently estimated using the linear law of redshifts. That idea has been advocated by H. Arp, C. Fulton, D. Carosati,[104] and some other astrophysicists.

Hubble's law of redshifts states that there is a *linear* galaxy redshift – distance relationship.[113] The same has been suggested earlier in this section. Examination of photons Ph.1 and Ph.2 emitted at point C in Fig. 22 also leads to the prediction of a linear redshift – distance relationship. That is for the longer the photon travels the more energy U it loses in direct proportion to the length of the travelled pathway as gradient D has a constant magnitude.

Nonetheless, a starlight-beam running through space may pass through one or more cosmic voids before reaching an observer. In the G-P-FS of a cosmic void, the photon doesn't lose energy since there are no gravitational interactions between the photon and other objects. As a result, the redshift – distance relationship is expected to be *nonlinear* at large cosmic distances, with redshift increasing at a slower rate than distance: At distances larger than the distance to an observer's G-P-FH, a deviation from Hubble's law of redshifts is predicted because, over large cosmic distances, the photon runs through more gravitational-potential-free-space the farther it travels. That is in agreement with the observed redshift–distance relationships reported by S. Perlmutter et al.,[114] S. van den Bergh,[115] and others.

69 The life of a star

In an-infinite-in-time universe, for each star that is born, another star has to die. As stars comprise entirely energy U, a typical star must emit, on the average, more energy U than it absorbs such that some energy is left for the formation of new stars.[x] This section presents a speculative description of the life of a star in an-infinite-in-time universe, in which the range of gravity is finite such that gravitational-potential-free space (G-P-FS) can exist. It has to be speculative because according to the current interpretations of astrophysical observations, there are no stars older than about 13.7 Gyr in the Milky Way galaxy, and in more remote regions of the entire universe.

It was suggested in section 43 that the photon in G-P-FS—that is, in a cosmic void according to the suggestion made in section 61—can travel at any speed. No forces are exerted on the photon that enters G-P-FS. The speed of the photon in a cosmic void can become less than the speed of light c as a result of the collisions with other photons and particles. It follows that material energy U (matter) can accumulate in a cosmic void in the form of slow-moving energies U_{photon}. The accumulation of matter in a cosmic void cannot last forever. The accumulated matter is expected to become the basic ingredient used in the formation of new stars.

As the photons in a G-P-FS aren't subject to any forces, they can bind upon a collision, and highly energetic photons can then be created. Eventually, the energies of some of the photons can become sufficiently high for the creation of U_{solid}-core particles,

---

[x] In the standard cosmological model, stars are assumed to be born primarily out of cosmic gas and dust generated in supernova explosions. In the infinite-in-time universe, however, the scenario in which supernovae generate enough gas and dust for the formation of all stars that must be born isn't realistic. Each supernova would have to lead to the formation of a sufficiently massive star that would later result in another supernova. Otherwise, given infinite time, there would be no more supernovae. The number of the supernovae and the amount of gas and dust generated in those explosions would decrease with time, as some of the gas and dust would be used up in the formation of stars having masses smaller than those of pre-supernova-size stars.

just like in the hypothetical two-gamma-photons collision. Thus, cosmic gas and dust can be generated in conjunction with the creation of U_{solid}-core particles, primarily out of energy U_{photon}. New stars and galaxies can then be born, the G-P fields of which will eventually connect to the galaxy filament. A high rate of cold-gas accretion inside cosmic voids was discovered in 2004, and justifiably called a "most tantalizing finding" (R. van de Weygaert et al.[116]). The birth of a new star in the G-P-FS of a cosmic void is consistent with the "high rate of cold-gas accretion" discovery.

Other astronomic observations appear to support the scenario of star and galaxy formation in cosmic voids. First, the majority of galaxies inside cosmic voids are relatively young—that is, their spectra are "bluer" than the spectra of typical filament galaxies. That was reported on, for example, by Rojas et al.[117], Kreckel et al.[118], and it supports the scenario of galaxy formation in cosmic voids. Second, if formed in a G-P-FS by the accretion of energy U at extremely low rates, galaxies would be expected to acquire at birth, and then maintain regular shapes that could be spiral or elliptical (i.e., disk-like) owing to the "self-contained" gravitation and the initial lack of interaction with other galaxies. That idea is supported by observations, which show no presence of irregular galaxies in cosmic voids (Goldberg et al.[119]). Third, as already mentioned, the majority of void galaxies are rich in cosmic gas (e.g., Kreckel et al.), which is consistent with the scenario of the high rate of cosmic gas creation in cosmic voids.

Furthermore, astronomic observations show that the rate of star formation is higher, while the fraction of passive galaxies is lower in cosmic voids comparing to the galaxy filament (E. Ricciardelli et al.[120]). Both those observations are consistent with the proposed cosmological model, in which cosmic gas—the basic ingredient used in the creation of stars and galaxies—is generated primarily inside cosmic voids.

The scenario of the birth of a star outlined above appears to be straightforward, at least at the conceptual level. To the contrary, the death of a star scenario cannot be straightforward because it requires the cores of U_{solid} particles (the electrons, the protons, the neutrons) to disperse in the form of material energy U_{fluid} such that a star can disappear ("evaporate") entirely. This means that the particles a star comprises would be expected to lose their energy

U_{solid} at such miniscule rates that those cannot be observed in any time frame available to a human observer. Consequently, the idea of eventual evaporation of the cores of U_{solid} particles (of U_{solid} energy) has to be considered speculative at this time, as pointed out at the beginning of this section.

Yet, from the perspective of the universe that is infinite in time, the evaporation of the entire energy U_{solid} that a star comprises can be expected because the SLT is an absolute law of nature. If it isn't absolute and energy U_{soild} doesn't evaporate such that a star eventually disappears, the universe would have to comprise an infinite number of stars that stopped evaporating energy, which isn't the case. An unavoidable conclusion of this suggestion is that some of the electrons and the protons that form the dark matter must comprise lesser amount of U_{solid} energies than the electrons and the protons that form the visible stars. Of course, that is a very strange and controversial conclusion. One mustn't forget though that we have plenty of experience with stars being born and stars that are still young, with no experience with stars that are dying by evaporation of matter.

At present, we don't have observational data required to study stars older than about 13-14 Gyr. If the universe is infinite in time, and if the sun is about 5 Gyr old, then it is rational to expect that it will be hundreds (perhaps tens of thousands) of Gyr before the sun dies by emitting all the energy U that forms the cores of the U_{solid} particles the sun comprises at this time, and those that still will be formed inside the core of the sun in the future.[y]

Direct observation of the loss of mass from U_{solid} particles in the time interval available to a human observer is most likely impossible. It is worth noting, however, that miniscule changes in the proton-to-electron mass ratio—that may have occurred in the time period of roughly half the age of the big-bang's universe—have been considered and investigated (e.g., E. Reinhold et al.[121]). However, the results of those investigations are inconclusive (e.g., R. I. Thompson et al.[122], N. Kanekar at al.[123]). It appears that the only substantiated argument in favor of the above death-of-a-star scenario could be the expectation that the U_{solid} particle cores

---

[y] If the sun was born out of cosmic gas and dust generated in a cosmic void as discussed earlier in this section, it would be at this time much older than 5 Gyr for the process of gas and dust formation would take a very long time.

should totally disperse in the very long time because the SLT is an absolute law of nature.

70 Black holes

If a black hole is thought of to be a singularity in physics terms, then the existence of a black hole is doubtful. That is because in the current physics, the highest known density of matter is the density of the U_{solid} cores of particles, of the electron and the proton, which would be much less than the matter density needed to form a tangible singularity. Another problem with black holes is Hawking radiation by means of which black holes evaporate, and eventually disappear. In physics terms, Hawking radiation isn't possible for nature cannot spontaneously execute her laws to create new particles out of energy U, which would lead to a greater degree of nonequilibrium. That is because the preferred state of matter is an equilibrium state (Law I, section 3). Note that the creation of the photon discussed in section 22 is the effect of nature executing spontaneous work to eradicate nonequilibrium in a G-P field.

Regardless, black holes, if those do exist in physical sense, must entirely evaporate in an infinite-in-time universe, just as stars have to entirely evaporate. Therefore, the concept of a black hole is acceptable from the perspective of the EE cosmological model, if a black hole is defined as the G-P field of a g-object from which no energy in the form of photons (U_{photon}) is released. At first glance, an emitted photon eventually exits G-P field of the photon-emitting object as illustrated in Fig. 21. However, that doesn't have to be always the case. Suppose that object m_0 is so massive that the entire energy of the emitted photon is released prior to the photon reaching the boundary of the G-P field of the g-object. Energy U_{photon} "dissolves" into energy U_{fluid}. The photon ceases to exist as a result of gravitational redshift that causes the release of photon's entire material energy U_{photon}. The object m_0 system becomes a “black hole” in the sense that no thermal radiation in the form of emitted photons can escape it. However, energy U has to be released from the U_{fluid} "atmosphere" (from the G-P field of the object) for there is an excess energy due to the dissolution of the photon. This is the result of the execution of the SLT by nature.

Without that release, there would be an infinite number of black holes in the universe, which isn't the case.

71 Cosmic microwave background

The discussion presented in this section isn't intended to criticize in any way the concept of cosmic microwave background (CMB) being a relic radiation from the big-bang event. The purpose of the discussion is to reason that an origin of CMB other than the big-bang's photon decoupling may present a rational alternative.

The observed CMB radiation appears to provide an argument that can be used in support of the concept of an-infinite-in-time universe. In such a universe, the abundancy of VO&VC stars and galaxies would be expected (section 66). On a large scale, those would be distributed homogeneously throughout the universe. CMB would be expected to exist as the ordinary thermal radiation emitted by ordinary VO&VC matter at very low frequencies from all regions of the universe.

Perhaps next to the Doppler-shift interpretation of cosmological redshifts, CMB being a relic radiation from the big-bang is widely considered to be the most important evidence supporting the big-bang model. Yet, the merits of that proposal are often questioned (A. K. T. Assis and M. C. D. Neves[124]). G. Burbidge[125] specifically points out that the big-bang model cannot adequately explain CMB. Regardless of those polemics, CMB being a relic radiation from the big bang appears to be just an unproven possibility.

In the EE cosmology, the amount of ordinary ("baryonic") dark matter in the universe has to be much larger than the amount of matter estimated from direct astronomic observations (recall the Stefan-Boltzmann law argument from section 66). Said another way, the great majority of galaxies are expected to be VO&VC galaxies. Those invisible to human eye galaxies are expected to be abundant deep in intergalactic space, and to form the vast majority of the star horizon of any observer. An unavoidable weakness of cosmological research is that little observations of gravitational interactions of dark matter deep in intergalactic space have been made, except for the observations of some gravitational lensing.

The invisible VO&VC galaxies and the associated CMB appear to present physically comprehensible candidates for dark matter and the radiation of dark matter, respectively. While, at this time, it is difficult to prove CMB to be radiation emitted by VO&VC galaxies, it needs to be realized that, in principle, it is possible to obtain such a proof.[z] It could be obtained by sending a cosmic probe with the purpose of detecting and mapping VO&VC objects deep in intergalactic space. Of course, that will not happen any time soon. Nonetheless, further improvements to the resolution of radio telescopes used to detect and map CMB and other radiation might, in the foreseeable future, let us detect ("see") individual VO&VC galaxies that emit CMB.

In claiming CMB to be a relic radiation, it is rarely mentioned that CMB represents just the most dominant part of cosmic background radiation. In the EE cosmological model, invisible galaxies that emit radiation at photon energies other than the CMB energies have to exist. In this regard, the observations show that "invisible" radiation beams other than CMB or visible light beams do exist. Those are much less frequent beams than CMB beams, as expected from the consideration of the Stefan-Boltzmann law. The existence of cosmic infrared and X-ray backgrounds is well documented.

The big-bang model implicitly rejects the possibility of the existence of stars/galaxies emitting radiation at the microwave frequencies because no star/galaxy would have time to cool to the microwave level, say, to 2.7 K as there has been roughly 13.7 Gyr only available for cooling. It appears that this rejection isn't based on any evidence. It is likely based on the need to sustain the viability of the big-bang model.

Say, it is year 1965 and Arno Penzias and Robert Wilson just discovered CMB. The standard cosmological model of the day is Einstein's model of the quasi-static universe, with cosmological constant included. (George Lemaitre hadn't publish his primeval atom proposal.) The question is: What would be the response of the scientific community to Penzias-Wilson's discovery? I believe good chances are that the response would be: "There must be a lot of invisible (VO&VC) galaxies out there. Those apparently are uniformly distributed throughout the universe, emitting the newly-

---

[z] In standard cosmology, it is assumed that dark matter doesn't emit radiation, which is based on the assertion that we don't detect dark matter's radiation. As far as I know, it is the only background to that assumption.

discovered CMB." I doubt that the response would be: "There had to be a photon decoupling about 13.7 billion years ago."

If CMB is indeed thermal radiation emitted by VO&VC stars, then small variations in the frequencies of the CMB photons, as observed, would be expected. That is for the CMB photons would be arriving at our observatories in the form of radiation emitted by stars having somewhat different temperatures, different masses, and located at different distances from the earth. Both different temperatures and masses, and different distances would affect the redshifts of the CMB beams. However, since the energies of CMB radiation beams at the time of emission would be very low since the surface temperatures of the emitting stars would be very low, the observed differences in the frequencies of CMB beams would be expected very small, which is the case.

72 Perpetuum mobile

A simple definition of perpetual motion is a motion that continues indefinitely without any input of external work. If one thinks of an infinite-in-time universe as a machine, one will soon realize that it is the only perpetual-motion machine that can exist, and it does according to the EE cosmological model. That is for nature has no hidden sources of external work. And no matter what the friction losses are, those will always be absorbed in conjunction with a contra work done by nature for she has no hidden storage bins to store those losses. It follows that an-infinite-in-time universe can be in motion indefinitely indeed.

73 Summary

According to the EE theory put forward in this book, dehumanized nature—which corresponds to physics at the fundamental level—is extremely simple. It comprises material energy *U* (matter), the second law of thermodynamics, the energy-equilibrium law, the material-energy-conservation law with the consequential nature's work and motion-conservation laws, and the action-reaction law. In the EE theory, there is no room for nature owning computers, thermometers, rulers, clocks, electrometers, etc. There is no room for unbalanced and fictitious forces, and abstract spacetime. And, there is no room for space empty of physically-real matter. The weakest point in the proposition of the simplicity of fundamental physics appears to be the spin-charge connection. If charge can be explained in terms of spin, as tentatively considered in chapter VI, with the heat and temperature measurement explainable in terms of material energy U and its density, the fundamental dimensions of physical quantities would be reduced to the three highly rational dimensions: mass (amount of material energy), length, and time.

REFERENCES

[1] R. P. Feynman (1963), *The Feynman Lectures on Physics*. Vol. I, Ch. 4-1 Addison-Wesley Publishing Co. (1997).
[2] I. Newton (1687), *Philosophiae Naturalis Principia Mathematica* p. 506 (Daniel Adee, New York 1848).
[3] M. B. Szymanski, Gravitation Photons Universe, Phys. Essays **23** 388 (2010).
[4] M. B. Szymanski, The physics of infinite in time universe, Phys. Essays **25** 455 (2012).
[5] M. B. Szymanski, The simple nature of physics, Phys. Essays **25** 590 (2012).
[6] M. B. Szymanski, The constancy of the speed of light, Phys. Essays **31** 89 (2018).
[7] S. Weinberg, *Dreams of a Final Theory*, Ch. VII. New York: Pantheon (1992).
[8] S. Hawking, *The grand design* (Bantam Books, New York 2010).
[9] L. Smolin, *How to Understand the Universe When You're Stuck Inside of It.* Interview by A. Gefter (2019)
[10] C. Rovelli, "Physics Needs Philosophy. Philosophy Needs Physics," Found. Phys. **48**, 481 (2018).
[11] Aristotle, *Physics* BK4 §11, (circa 379 B.C.) Printed by Princeton University Press (1984).
[12] J. B. Barbour, "The Emergence of Time and Its Arrow from Timelessness," *Physical Origins of Time Asymmetry*, eds. J. J. Halliwell, J. Perez-Mercader & W. H. Zurek 405, Cambridge University Press (1995).
[13] C. Rovelli, Forget time, Foundations of Physics **41** (9) 1475-1490 (2009).
[14] L. Smolin, *Time Reborn* p. 247, (Random House of Canada Limited, 2013).
[15] I. Newton (1687), *Philosophiae Naturalis Principia Mathematica* (Daniel Adee, New York 1848).
[16] A. Einstein, "The Foundations of the General Theory of Relativity." Annalen der Physik **49** 769 (1916). English translation in *The Principle of Relativity* (Dover Publications, 1952).
[17] T. Kaluza, "The unity problem in physics," Sitzungsber, Preuss. Akad. Wiss. Berlin. 966 (1921).
[18] O. Klein, "Quantum Theory and Five-Dimensional Theory of Relativity." Zeitschrift für Physik A, **37** 895 (1926).
[19] C. Brans, R. H. Dicke, "Mach's Principle and a Relativistic Theory of Gravitation." Phys. Rev. 124, 925 (1961).
[20] J. Moffat, *Reinventing gravity* (Thomas Allen Publications, Toronto 2008).
[21] M. Milgrom, "A Modification of the Newtonian Dynamics as a Possible Alternative to the Hidden Mass Hypothesis," Astroph. J. **270**, 365-370 (1983).
[22] J. D. Bekenstein, "Relativistic gravitation theory for the modified Newtonian dynamics paradigm," Phys. Rev. D **70** 083509 (2004).
[23] P. G. O. Freund, A. Maheshwari and E. Shonberg, "Finite-Range Gravitation," Astroph. J. **157**, 857-967 (1969).
[24] D. G. Boulware, S. Deser, "Can Gravitation Have a Finite Range?" Phys. Rev. D **6** 3368 (1972).

[25] S. V. Babak, L. P. Grishchuk, " Finite-Range Gravity and Its Role in Gravitational Waves, Black Holes and Cosmology," Int. J. Mod. Phys. D **12** 1905-1959 (2003).
[26] A. D. Allen, "Finite Gravity: From the Big Bang to Dark Matter," Astronomy and Astrophysics **3** 180 (2013).
[27] S. Bose, *et al.*, "Spin Entanglement Witness for Quantum Gravity'" Phys. Rev. Lett. **119**, 240401 (2017).
[28] C. Marletto, V. Vedral, "Gravitationally induced entanglement between two massive particles is sufficient evidence of quantum effects in gravity," Phys. Rev. Lett. **119**, 240402 (2017).
[29] E. Verlinde, "On the origin of gravity and the laws of Newton," J. High Energ. Phys. **2011**, 29 (2011).
[30] S, M. Carroll, R. N. Remmen, " What is the entropy in entropic gravity?" Phys. Rev. D, **93**, 124052 (2016).
[31] T. A. Wagner, *et al.*, "Torsion-balance tests of the weak equivalence principle," Class. Quantum Grav. **29** 184002 (2012).
[32] P. Touboul, *et al.*, "MICROSCOPE Mission: First Results of a Space Test of the Equivalence Principle," Phys. Rev. Lett. **119**, 231101 (2017).
[33] J. Q. Quach, "Gravitational Casimir Effect," Phys. Rev. Lett. **114**, 081104 (2015).
[34] R. A. Norte, "Platform for Measurements of the Casimir Force between Two Superconductors," Phys. Rev. Lett. **121**, 030405 (2018).
[35] B. P. Abbott *et al.*, GW170817: Observation of Gravitational Waves from a Binary Neutron Star Inspiral, Phys. Rev. Lett. **116** 061102 (2016).
[36] I. Newton, Letter to Richard Bentley. Feb. 25 (1692).
[37] Lee, J. G. *et al.*, "New Test of the Gravitational 1/r2 Law at Separations down to 52 μm," Phys. Rev. Lett. **124**, 101101 (2020).
[38] A. Einstein (1920), Relativity: *The Special and General Theory*, The proof 97, Penguin Books 2006.
[39] D. W. Sciama, "On the Origin of Inertia," MNRAS **113**, 34 (1953).
[40] E. Mach (1893), *The Science of Mechanics*. Chapter II. Section VI. The Open Court Publishing Co. (1919).
[41] A. Einstein, "The Foundations of the General Theory of Relativity." Annalen der Physik **49** 769 (1916). English translation in *The Principle of Relativity* (Dover Publications, 1952).
[42] A. Einstein (1922), *The Meaning of Relativity*. Princeton University Press (1956).
[43] H. Poincaré, Note de "Sur la dynamique de l'électron," C.R. T.140 (1905) 1504-1508.
[44] A. Einstein (1918), "Über Gravitationswellen," Königlich Preußische Akademie der Wissenschaften (Berlin). Sitzungsberichte,154–167.
[45] B. P. Abbott *et al.*, GW170817: Observation of Gravitational Waves from a Binary Neutron Star Inspiral, Phys. Rev. Lett. **116** 061102 (2016).
[46] R. P. Feynman (1963), *The Feynman Lectures on Physics*. Vol. I, Ch. 13. Addison-Wesley Publishing Co. (1997).
[47] C. W. Misner, K. S. Thorne, J. A. Wheeler, *Gravitation* §20.4. W. H. Freeman and Company, San Francisco (1973).

[48] B. P. Abbott *et al.*, GW170817: Observation of Gravitational Waves from a Binary Neutron Star Inspiral, Phys. Rev. Lett. **116** 061102 (2016).
[49] J. H. Taylor, J, M. Weisberg, "A new test of general relativity: gravitational radiation and the binary pulsar PSR 1913+16," Astroph. J. **253**, 908 (1982).
[50] C. W. Misner, K. S. Thorne, J. A. Wheeler, *Gravitation* §20.4. W. H. Freeman and Company, San Francisco (1973)
[51] J. H. Taylor, J, M. Weisberg, "A new test of general relativity: gravitational radiation and the binary pulsar PSR 1913+16," Astroph. J. **253**, 908 (1982).
[52] F. Zwicky, The Redshift of Extragalactic Nebulae, Helvetica Physica Acta, Vol. 6, p. 110-127, 1933
[53] R. V. Pound, G. R. Rebka, Gravitational Red-shift in Nuclear Resonance. Phys. Rev. Lett. **9** 439 (1959).
[54] N. L. S. Carnot (1824), *Reflection on the Motive Power of Heat*. John Wiley & Sons, London (1897).
[55] W. Thomson, *Carnot's Theory of the Motive Power of Heat*. Reprinted in John Wiley & Sons (1897).
[56] J. C. Maxwell, Theory of Heat. 3rd. Edition. Longmans, Green and Co., London (1972).
[57] A. Einstein, Jahrbuch der Radioaktivitat und Elektronik 4, 411 (1907). (English translation, in "On the relativity principle and the conclusions drawn from it", in "The Collected Papers", v.2, Doc. 47.)
[58] I. Newton (1687), *Philosophiae Naturalis Principia Mathematica* p. 83 (Daniel Adee, New York 1848).
[59] J. A. Klimchuk Solar physics **41,** 234 (2006)
[60] M. Aschwanden, *Physics of the Solar Corona: An Introduction with Problems and Solutions*. (Springer 2005)
[61] R. P. Feynman (1963), *The Feynman Lectures on Physics*. Vol. III, Ch. 1. Addison-Wesley Publishing Co. (1977)
[62] A. *Einstein,* Lichtgeschwindigkeit und Statik des Gravitationsfeldes (The speed of light and the statics of the gravitational field), *Annalen der Physik.* ***38*** *355* (1912).
[63] R. H. Dicke, Gravitation without a Principle of Equivalence, Reviews of Modern Physics, **29** (3) 363 (1957).
[64] J. W. Moffat, Superluminary Universe: A Possible Solution to the Initial Value Problem in Cosmology, Int. Journal of Modern Physics D, **02** (03) 351 (1993).
[65] A. Albrecht and J. Magueijo, Time varying speed of light as a solution to cosmological puzzles , Phys. Rev. D **59** 043516 (1999).
[66] P. Dirac, Is there an Aaether, Nature, **168** 906 (1951).
[67] J.-P. Vigier, Interactions of Internal Inertial and Phase Space Motions of Extended Particle Elements Moving in Dirac's Real "Aaether" Model, in Editor Valeri V. Dvoeglazov, *Einstein and Poincare: The Physical Vacuum*, 1 C. (Roy Keys Inc., 2006).
[68] J. Levy, Is the Aaether Entrained by the Motion of Celestial Bodies? What do the Experiments Tell Us? arXiv:1204.1885v3 (2013).
[69] M. C. Duffy, The Aether Concept in Modern Physics, in Editor Valeri V. Dvoeglazov, *Einstein and Poincare: The Physical Vacuum*, 1 C. (Roy Keys Inc., 2006).

[70] R. P. Feynman, QED, the strange theory of light and matter, p.15 (Princeton University Press, 1985).
[71] A. Michelson & E. Morley, On the Relative Motion of the Earth and the Luminiferous Aether, American Journal of Science **34** 333 (1887).
[72] George G. Stokes, On the Aberration of Light, Philosophical Magazine **27** 177 (1845).
[73] A. Einstein, On the Electrodynamics of Moving Bodies, Annalen der Physik **322** 891 (1905). English translation in *The Principle of Relativity* (Dover Publications, 1952).
[74] A. Einstein, "Does the Inertia of a Body depend on its Energy Content?" Annalen der Physik **18**, 639 (1905).
[75] A. Einstein, "The Foundations of the General Theory of Relativity." Annalen der Physik **49** 769 (1916). English translation in *The Principle of Relativity* (Dover Publications, 1952).
[76] H.A. Lorentz, Michelson's Interference Experiment (1895). English translation in *The Principle of Relativity* (Dover Publications, 1952).
[77] H. Minkowski, Address delivered at the Eightieth Assembly of German Natural Scientists and Physicians (1908). English translation in *The Principle of Relativity* (Dover Publications, 1952).
[78] A. Einstein, "Does the Inertia of a Body depend on its Energy Content?" Annalen der Physik **18**, 639 (1905).
[79] M. Joeveer, J. Einasto & E. Tago, Spatial distribution of galaxies and of clusters of galaxies in the southern galactic hemisphere, MNRAS **185** 357 (1978).
[80] S. A. Gregory & L. A. Thompson, The Coma/A1367 supercluster and its environs, Astrophysical Journal **222** 784 (1978).
[81] R.P. Kirshner, A. Oemler Jr., P.L. Schechter & S.A. Schectman, Astrophysical Journal **248** L57 (1981).
[82] J. Huchra, M. Davis, D. Latham & J. Tonry, Astrophysical Journal, Supplement **52** 89 (1983).
[83] J. R. Bond, L. Kofman & D. Pogosyan, How Filaments Are Woven into the Cosmic Web, Nature **380** 603 (1966).
[84] Rien van de Weygaert & Erwin Platen, Cosmic Voids: Structure, Dynamics and Galaxies, Int. Journal of Modern Physics: Conference Series **1** 41 (2011).
[85] B. R. Granett, M. C. Neyrinck, I. Szapudi, An Imprint of Superstructures on the Microwave Background Due to the Integrated Sachs-Wolfe Effect, Astrophysical Journal **683** L99-L102 (2008).
[86] C. Foster & N. Lorne, The Size, Shape, and Orientation of Cosmological Voids in the Sloan Digital Sky Survey, Astrophysical Journal **699** 1252 (2009).
[87] D. Pan, M. S. Vogeley, F. Hoyle, Y.-Y. Choi & C. Park, Cosmic Voids in SDSS Data Release 7, arXiv:1103.4156v2 (2011).
[88] F. Zwicky, The Redshift of Extragalactic Nebulae, Helvetica Physica Acta, **6** 110 (1933).
[89] P. A. M. de Theije, P. Kartgert & E. van Kampen, The Shapes of Galaxy Clusters, Mon. Not. R. Astron. Soc. **273** 30 (1995).
[90] J. Bahcall, T. Piran & S. Weinberg, Dark Matter in the Universe, (World Scientific, 2004).

[91] N. A. Bahcall & A. Kulier, Tracing mass and light in the Universe: where is the dark matter?, arXiv:1310.0022v2 [astro-ph.CO] (2014).
[92] J. Bennet, M. Donahue, N. Schneider & M. Voit, The Essential Cosmic Perspective, 447-448 (Pearson Addison-Wesley, 2007).
[93] X.-X. Xue, H.-W. Rix, P. R. Fiorentin, T. Naab, M. Steinmetz, C. van den Bosh, T. C. Beers, Y.S. Lee, E. F. Bell, C. Rockosi, B. Yanny, H. Newberg, R. Wilhelm, S. Kang, C. Smith, and D. P. Schneider, The Milky Way's Circular Velocity Curve to 60 kpc and an Estimate of the Dark Matter Halo Mass from Kinematics of ~2400 SDSS Blue Horizontal Branch Stars, Astrophys. J. **684** 1143 (2008)
[94] E. Hubble, A Relation between Distance and Radial Velocity among Extra-galactic Nebulae, Proc. N. A. S. **15** 168 (1929).
[95] E. Hubble, The Observational Approach to Cosmology, 9 (Clarendon Press 1937).
[96] R. Penrose, The Road to Reality, 462 (Random House: London, 2004).
[97] F. Wilczek, The Lightness of Being, 106 (Basic Books, 2008).
[98] V. M. Slipher, The Radial Velocity of the Andromeda Nebula, Lowell Observatory Bulletin **58** 56 (1913).
[99] E. Hubble, The Observational Approach to Cosmology, 26 (Clarendon Press 1937).
[100] Harvard web page: www.cfa.harvard.edu/seuforum/galSpeed/.
[101] F. Zwicky, On the Red Shift of Spectral Lines Through Interstellar Space, Proceedings of the National Academy of Sciences **15** 773 (October 1929).
[102] P. A. Violette, Is the Universe Really Expanding, Astrophysical Journal **301** 544 (1986).
[103] J. C. Pecker, A. P. Roberts & J. P. Vigier, Non-velocity red shifts and photon-photon interactions, Nature **237** 227 (1972).
[104] J. C. Pecker & J. P. Vigier, A possible tired-light mechanism, Apeiron **2** 19 (1988).
[105] M. Harney, The Cosmological-Redshift Explained by the Intersection of Hubble Spheres, Apeiron **2** 288 (2006).
[106] J. F. G. Julia, Simple Considerations on the Cosmological Redshift, Apeiron **15** 325 (2008).
[107] A. Gosh, Velocity-Dependent Inertial Induction, Apeiron **9–10** 95 (1991).
[108] L. Marmet, On the Interpretation of Spectral Red-Shift in Astrophysics: A Survey of Red-Shift Mechanisms - II, arXiv:1801.07582v1.
[109] H. Arp, Evolution of Quasars into Galaxies and Its Implications for the Birth and Evolution of Matter, Apeiron **5** 135 (1998).
[110] J. V. Narlikar, Noncosmological redshifts, in Quasars, IAU Symp., Reidel 463 (1986).
[111] H. Arp, C. Fulton, D. Carosati, Intrinsic Redshifts in Quasars and Galaxies (http://www.haltonarp.com/articles/intrinsic_redshifts_in_quasars_and_galaxies.pdf).
[112] J.-C. Pecker, Local abnormal redshifts, 'Current Issues in Cosmology' eds. Jean-Claude Pecker and Jayant V. Narlikar, 217 (Cambridge University Press 2006).
[113] E. Hubble, The Law of Red-shifts MNRAS **113**, 658 (1953)
[114] S. Perlmutter, G. Aldering, G. Goldhaber, R.A. Knop, P. Nugent, P. G. Castro, S. Deustua, S. Fabbro, A. Goobar, D. E. Groom, I. M. Hook, A. G.

Kim, M. Y. Kim, J. C. Lee, N. J. Nunes, R. Pain, C. R. Pennypacker, R. Quimby, Measurements of Ω and Λ from 42 High-Redshift Supernovae, Astrophysical Journal **517** 565 (1999).
[115] S. van den Bergh, Is the velocity-distance for galaxies linear?, Proc. Natl. Acad. Sci. **90** 4793 (1993).
[116] R. van de Weygaert, K. Kreckel, E. Platen, B. Beygu, J. H. van Gorkom, J. M. van der Hulst, M. A. Arag´on-Calvo, P. J. E. Peebles, T. Jarrett, G. Rhee, K. Kovaˇc, C.-W. Yip, The Void Galaxy Survey, arXiv:1101.4187v1 (2011).
[117] R. R. Rojas, M. S. Vogeley, F. Hoyle, J. Brinkmann, Photometric Properties of Void Galaxies in the Sloan Digital Sky Survey, arXiv:astro-ph/0307274v2 (2004).
[118] K. Kreckel, E. Platen, M.A. Aragon-Calve, J.H. van Gorkom, R. van de Weygaert, J. M. van der Hult & B. Beygu, The Void Galaxy Survey: Optical Properties and HI Morphology and Kinematics, The Astronomical Journal, **144:16**, (2012).
[119] D. M. Goldberg, T. D. Jones, F. Hoyle, R. R. Rojas, M. S. Vogeley, M. R. Blanton, The Mass Function of Void Galaxies in the SDSS Data Release 2, Astrophysical Journal **621** 643 (2005).
[120] E. Ricciardelli, A. Cava, J. Varela & V. Quilis, The star formation activity in cosmic voids, MNRAS **445** 4045 (2014).
[121] E. Reinhold, R. Buning, U. Hollenstein, A. Ivanchik, P. Petitjean, W. Ubachs, Indication of a Cosmological Variation of the Proton-Electron Mass Ratio Based on Laboratory Measurement and Reanalysis of H2 Spectra, Phys. Rev. Lett. **96** 151101 (2006).
[122] R. I. Thompson, J. Bechtold, J. H. Black, D. Eisenstein, X. Fan, R. C. Kennicutt, C. Martins, J. X. Prochaska, Y. L. Shirley, An Observational Determination to the Proton to Electron Mass Ratio in the Early Universe, Astrophysical Journal **703** 1648 (2009).
[123] N. Kanekar, W. Ubachs, K. M. Menten, J. Bagdonaite, A. C. Brunthaler, S. Henkel, S. Muller, H. L. Bethlem, M. Dapra, Constraints on changes in the proton-electron mass ratio using methanol lines, MNRAS: Letters **448** L104-L108 (2015).
[124] A. K. T. Assis & M. C. D. Neves, History of the 2.7 K Temperature Prior to Penzias and Wilson, Apeiron **2** (3) (1995).
[125] G. Burbidge, The Stage of Cosmology, in 'Current Issues in Cosmology' edited by Jean-Claude Pecker and Jayant V. Narlikar, 3 (Cambridge University Press 2006).

Milton Keynes UK
Ingram Content Group UK Ltd.
UKHW021000220324
439787UK00001B/12

9 781916 849730